ADVANCES IN UNDERWATER TECHNOLOGY AND OFFSHORE ENGINEERING

Volume 1

Developments in Diving Technology

**ADVANCES IN
UNDERWATER TECHNOLOGY
AND OFFSHORE ENGINEERING**

Forthcoming Volumes:–

ADVANCES IN UNDERWATER TECHNOLOGY AND OFFSHORE ENGINEERING

Volume 1

Developments in Diving Technology

Proceedings of an international conference, (Divetech '84) organized by the Society for Underwater Technology, and held in London, UK, 14–15 November 1984

Published by
Graham & Trotman Ltd.

First published in 1985 by
Graham & Trotman Limited
Sterling House
66 Wilton Road
London SW1V 1DE

British Library Cataloguing in Publication Data

Divetech '84 *(1984: London)*
  Advances in underwater technology and offshore engineering:
proceedings of an international conference (Divetech '84), organised by
the Society for Underwater Technology and held in London, UK, 14-15
November 1984.
Vol. 1: Developments in diving technology
1. Ocean engineering
I. Title II. Society for Underwater Technology 627'.7 TC1505

Divetech '84 (1984: London)

ISBN-13: 978-94-010-8700-1   e-ISBN-13: 978-94-009-4970-6
DOI: 10.1007/978-94-009-4970-6

© Society for Underwater Technology
Softcover reprint of the hardcover 1st edition 1984

Typeset in Great Britain by Spire Print Services Ltd, Salisbury

Contents

Foreword

Up to about 30 years' ago diving activity was centred primarily on the naval services, who provided a lead in the development of equipment, techniques and procedures. Apart from one or two spectacular salvage undertakings, the main commercial activity up until that time was fairly low-key work in docks and harbours.

The concept of saturation diving emerged from subsea habitats of which Captain Cousteau was one of the pioneers. This led the way to commercial development in support of exploration and the production of offshore oil and gas, and I believe that my friend Henri Delauze was one of the first to mount the subsea habitat on deck and provide a sealed bell to convey divers from the habitat to the seabed without change of pressure.

A remarkable feature of offshore oil and gas technology in the North Sea has been the willingness of all concerned to exchange information regarding R&D. This has had a major effect on the advance in technology over the last few years. As far as diving is concerned, it is to some extent 'Hobson's Choice'. Legal patents are difficult to achieve in this field, and the casual nature of diver employment to date has meant that ideas and techniques circulate almost as freely as the divers themselves. In addition, the advertising of the new technologies which one has to offer almost automatically means disclosure of what otherwise might be secret.

Nevertheless, this exchange of information has been made considerably more effective by the conferences and seminars organized by the SUT, a society of which I had the honour to be President two years' ago. I am sure that this conference which is about to begin will prove as beneficial as those in the past. Its importance is illustrated by the fact that the Duke of Edinburgh has honoured us by the message published in the conference brochure.

This is a technical conference and no doubt the major interest will be centred on diving in deeper and deeper waters, hot tapping, cold tapping and the other activities which are on the present frontiers of science. However, I think it is important that we should not lose sight of the fact that the purpose of technology is to serve commerce. In this connection there are one or two items on the borderline between technology and commerce which I believe would merit more attention than they at present receive.

If one considers the enormous investment and deployment of skilled manpower involved in operating a DSV, it is surprising that in many cases this is to support only a single diver out of the bell. Existing technology can produce larger bells as well as two or more bells in a single vessel to enable four divers to be at work simultaneously out of the bell, but this requires the co-operation of the client to produce a suitable programme of work. In view of the enormous safety gain given by the provision of two bells, it is to be hoped that clients will 'bend' their programmes to make this practicable, although the commercial advantage by itself should be more than sufficient.

I believe that effort would then be well spent in reducing the time between when the divers' return to the bell at the end of one bell run and when they emerge at the beginning of the next. It may be that time and motion studies on this activity would produce quite substantial design changes in future installations. What I have particularly in mind is the provision of suitable space for the briefing of the new shift of divers by those who have just come up from the bottom.

Considerable effort is now being rightly directed to providing the diver tools and equipment to make better use of his working time. The present highly seasonal nature of diver employment with a sharp peak in June, seems to me to be very inefficient, and is only possible at present by reason of the surplus of underemployed vessels and men. It is to be hoped that better weather working capability will produce a situation in which regular employment throughout most of the year is the rule rather than the exception.

I would suggest that more originality should be directed to the design of diving systems for platform inspection. I envisage the possibility of a conventional fixed habitat and bell installation with galleries at various levels running round the platform. Here the divers could walk round with their heads in the dry as in a bell trunk, and tap in their umbilicals at various points.

November 1984   John Houlder

1

Comparison of Past with Present

James A. Lawrie, Humphreys & Glasgow Ltd.,
London

INTRODUCTION

With a title such as this there is a great temptation to venture back into the annals of history to the Greek breath-hold divers, William Phip's use of an open-bottom bell for salvage in 1687 to the more recent days of the invention of the hard hat and Haldane's pioneering work in stage decompression. However, as much as our industry owes to the pioneer salvors, builders and scientists of the distant past, the real history of the commercial subsea industry has been written over the past two decades. With minor exceptions our industry is built upon the development of offshore oil (Fig. 1), and although the first successful open water well was drilled in the Gulf of Mexico in 1947, the emergence of our industry really began in the early 1960s.

Within the constraints of this paper the author will endeavour to identify the major trends and developments in the undersea industry over the past twenty years which have laid the foundation of today's significant global industry. With few exceptions, 'firsts' will not be identified with individuals or companies since, as in many industries, successful work followed parallel paths in different parts of the world and often pioneering activities by some were commercialized by others. These points may be best argued by the old hands in the bar!

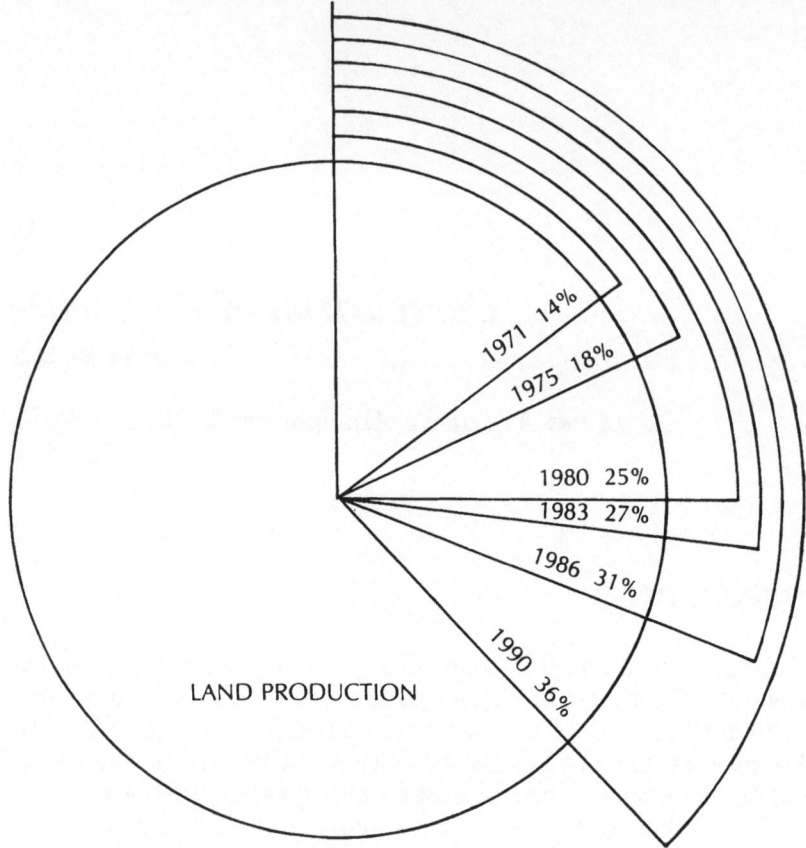

Fig. 1.1 Offshore free world petroleum production as a percentage of total

THE START OF THE BUSINESS

The origins of the American industry may be traced primarily to the early requirements of the offshore oil industry in Louisiana, Texas and California for limited subsea intervention. In Europe the foundations were established in scientific, military and inshore construction activities. The founders of the business began in many ways — abalone diving, pier and harbour work, marine salvage — and many had received their training from the world's navies. Few had the extensive formal educations that would be helpful in the future in managing the multimillion dollar enterprises which they would found, but all were entrepreneurs with strong personal drive, an abiding personal sensitivity for the subsea domain, and a sense of adventure. Most were undercapitalized in an industry that would ultimately be noted for its capital intensity, but with pick-up trucks, old air compressors, hard hats, early lightweight equipment, scuba gear, the odd decompression chamber, technology gleaned from the navy and little or no money in the bank, they set out to meet the needs of the offshore oil industry. Unlike the early aviation pioneers, many of the founders of our industry are still among us — Murray Black, Mike Hughes, Henri Delauze, Lad Handleman, Dan Wilson, Andre Galerne, Dick Evans — to name just some. A few companies, like Ocean Systems, were established by large corporations, but the vast majority were begun by individuals with the help of a few close colleagues.

THE SIXTIES

The offshore oil and gas industry was still in its infancy in the early sixties and the great technical developments in drilling rigs and production, which today are accepted practice, were just beginning to emerge. But in many areas of the world, offshore prospects posed better financial potential than those on land despite greater technical problems and greater costs. As the market for diving services continued to expand, particularly in the USA, the number of small companies proliferated. The first sorting out of the companies began around 1962 when various deeper operations dictated the use of heliox techniques. Only a few of the companies could meet this technical demand, and many of the others remain to this day as small air-diving-only local enterprises.

By 1966 saturation diving techniques, first developed outside the industry and subsequently refined within the industry, ushered in a new deep-water, extended-duration capability which was to have major impact particularly in construction operations. With support of deeper exploration drilling, saturation and protracted decompression came the emergence of the deep diving system — a very expensive tool. The need to invest hundreds of thousands of dollars and often millions in operating hardware caused marked changes in the industry. Many small and medium sized companies were acquired by larger corporations better able to provide the financial resources. Others merged to combine their strengths. Over a period of five years the foundations were laid for the evolution of 10 or 12 large international subsea contractors. The industry had entered a new phase, one in which a new group, accountants and financial controllers, fussed about receivables, cashflow, investment, depreciation and DCF to the eternal frustration of operationally-orientated management.

During the late sixties size too became important as the larger companies began to develop their business outside their home countries. Formation of legal entities, establishment of distant offices and operating bases, multicurrency trading, recruitment of foreign staff and all the other attendant problems in establishing a viable foreign presence necessitated resources which only the largest firms could muster. The move toward multinationalism was accelerated in 1969 with the discovery of Ekofisk oil which ushered in the great North Sea developments, and ultimately the largest and most demanding market for the subsea industry. But expansion was not restricted to the North Sea. The subsea industry followed the major oil companies throughout the world and gradually established themselves in the Mediterranean, Middle East, Africa, Asia and South America during the period.

THE SEVENTIES

During the latter half of the sixties and the first half of the seventies the subsea industry was in a continuous growth period matching the accelerating tempo of offshore exploration and development. The OPEC embargo of 1973 provided further impetus for the expanded offshore development as nations grew increasingly concerned about their dependence on foreign crude and oil prices rose four-fold in one year.

The financial well-being of the subsea industry and offshore activity level of the major oil companies have always been inexorably linked. From 1965 through 1975 profit margins were good; albeit, much of the profits were ploughed back into the companies to fund R & D and finance additional capital equipment. The year 1975 was unique in the annals of history of the industry for it was the only time that demand for services exceeded the industry's supply of trained personnel and equipment. In margin terms it was a period never to be repeated.

An interesting aspect of the business which evolved over that ten year period was that of supplying diving systems and other major capital equipment for offshore operations. In the mid-sixties there were no hardware suppliers who specialized in providing bells and similar equipment to the industry. The larger diving companies established their own design engineering groups, employed outside pressure vessel fabricators, and outfitted the equipment in their own shops. This approach persisted until 1975 by which time several supply companies had developed their own expertise, and the diving companies switched to procurement of standardized, off-the-shelf hardware.

During the first decade the majority of the revenues in the industry was derived from support of exploratory drilling activities. But during the period of 1972 and continuing into 1978 this area of the market was eclipsed by the construction sector. With numerous production facilities being installed in the North Sea, the Gulf of Mexico and elsewhere, those companies with large saturation facilities, substantial cadres of personnel and specialized knowledge in construction techniques overshadowed the multiservice line companies in both revenues and profits.

The year 1977 marked the beginning of the first 'great depression' in the subsea industry. For reasons which are still somewhat inexplicable, a major decline in exploratory drilling took place. This downturn coincided with the rundown in the first wave of construction of North Sea production facilities. Overexpansion and speculative investment in the subsea industry contributed to excess capacity in a suddenly shrinking market. Margins plummeted under the force of competitive pressures and revenues began to sink.

Over the three year period of 1978 through 1980, the subsea industry had far more capacity than could be marketed. Equipment was deployed at depreciation rates or less. Labour rates and wages stagnated. Mobilization fees were no longer charged and equipment was left on vessels on speculation . Several contractors went into bankruptcy and others were merged at prices less than the book value of their assets. Equipment crowded the operating bases, and the cost of depreciation of unutilized assets produced a flood of red ink on the ledgers.

The industry weathered the storm, often by borrowing to cover current losses, and in part mortgaged its future viability. As we all recognize, the oil industry is not simply driven by the need to continually find and produce new reserves; politics, taxes, war and revolution singly or in combination can and do exert more influence on the market place. Thus the overthrow of the Shah in 1979 and the subsequent cutback in Iranian oil production triggered major short-term shortages of fuel. Over the ensuing twelve months the price of crude more than doubled. The stage was set once again for a major expansion in offshore activity.

THE EIGHTIES

Throughout 1979 the subsea industry was still in the trough of depression. Several major operators were still paying the price for previous misadventures into ownership of light construction vessels, but rig utilization was on the upturn (Fig. 2) and there was light at the end of the tunnel. 1981 was a good year as excess capacity was gradually absorbed. The major oil companies were forecasting $50 and $70/barrel oil within a decade. The offshore industry became bullish, and rates for rigs more than doubled. The number of platforms installed reached a new

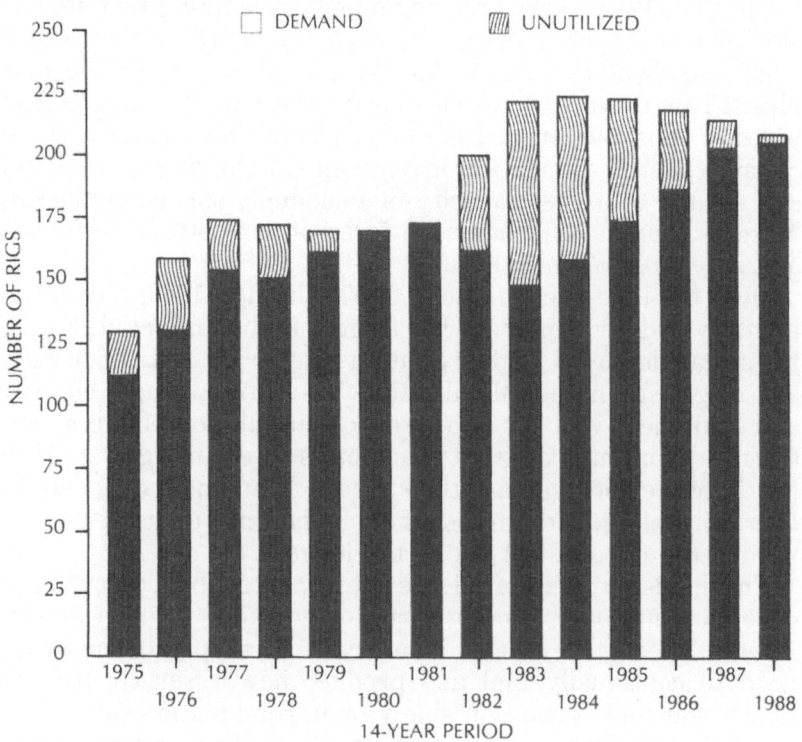

Fig. 1.2   Worldwide utilization of semisubmersibles/drillships

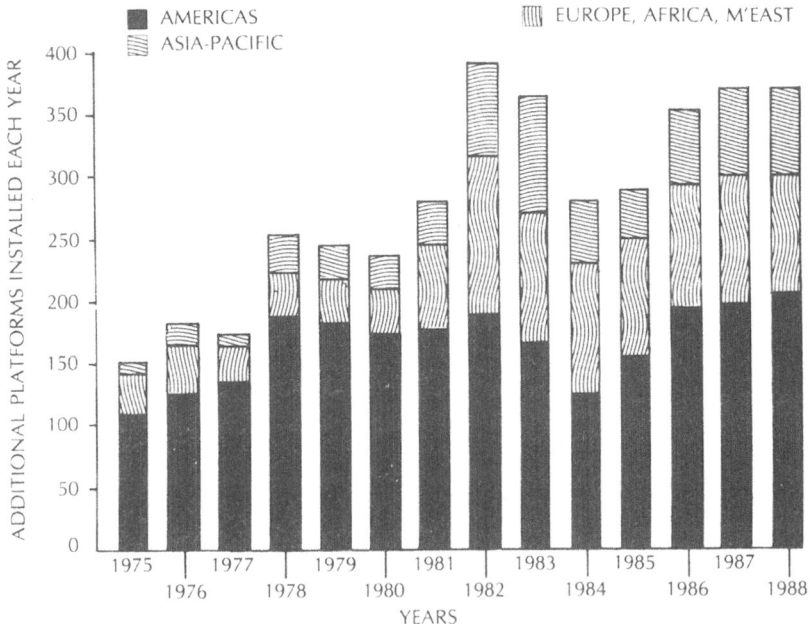

Fig. 1.3　Offshore production platforms by region

high (Fig. 3). The subsea industry was slower than most to recover and it was not until late 1981 that dayrates returned to 1975 levels. Even then, with continuously escalating costs over the intervening period, profit margins never again reached their former levels.

By 1983 the fundamentals of oil supply and demand took hold; the world was awash in a sea of surplus oil and gas. Conservation of energy as a result of higher cost had produced a more rapid and dramatic decline in consumption than had been anticipated. World recession further reduced commercial energy demand. Utilization of alternate fuel sources had increased. The oil price bubble burst and started a slow decline. Exploration and production budgets of the major oil companies were slashed, and once again the subsea industry was in trouble.

During the latter half of 1983 and 1984 we have witnessed a gradual return to normalcy in both the offshore and subsea industries. Crude prices have stabilized for the time being. Tax changes have opened up previously marginal fields for development. Oil companies, aside from buying each other, are giving more serious attention to restoration of depleted reserves. Con-

cern is being shown for declining domestic production and
potential foreign exchange problems from imports.

Although this bodes well for a period of stability in the market, it must be recognized that current and projected near-term
rates for undersea services are too low for the investment, sea-
sonality, and technological obsolescence inherent in the subsea
business. Profit margins are skimpy and are likely to remain so
for the forseeable future.

SUBSEA TECHNOLOGY

It has always been popular to compare the advances of manned
'outer space' with those of 'inner space'. The corollary is in some
ways valid, particularly with respect to man's penetration of an
unknown environment, but the problems to be overcome, the
technology that has been required, and the economic/political
driving forces have been quite different.

First of course are the differences in basic environment —
space with one bar negative pressure, extremes of cold and heat
and high G-forces during launch; subsea with many bars of positive pressure, moderate temperature differences but a high
thermal conductivity medium. The second major difference lies
in the basics of the astronaut being physically protected from
the environment and artificially provided a near normal life
medium (albeit analagous to the subsea one atmosphere system), versus the physiological adaptation required of the diver
as he transits the water column. Finally there is the difference
in fundamental drive and magnitude of resources applied to
penetration of the two frontiers. Manned space flight has been
as much a matter of political esteem of the USSR and USA as
scientific enquiry into the unknown. Hundreds of billions of dollars have been spent on manned space flight over the past two
decades, and the programme is only now entering a phase
wherein commercial potential may be realized. By contrast,
with the exception of some investment by the world's navies for
military objectives, developments in manned subsea intervention have been basically driven by commercial factors.

Building the Technical Foundation

As we review the short technical history of the subsea industry,
it becomes apparent that a relatively small group of technical

people with limited financial resources moved the industry forward quite rapidly. As the 1960s began, air diving and limited depth/duration heliox diving were the only forms of intervention available. The concepts and techniques for saturation diving, with the prospect of unlimited bottom time and greater depth penetration, were ultimately to have profound effect on the industry. The work in the early and mid-sixties of Bond, Link, Cousteau, Lambertsen and others broke down the depth and duration barriers.

Those new to the industry may not recall the controversies that existed at the time between those favouring bottom dwelling habitats versus those supporting the use of deck chambers for personnel storage. Commercially orientated people, however, tend to be practical realists, and the pattern of the commercial deep diving system, in various sizes and shapes, soon became the industry norm. It is noteworthy that this early pattern of bell and deck chamber, albeit with multitude of minor refinements, has persisted in our industry and over 450 such systems are in use today.

During the mid-1960s the aircraft and aerospace companies cast an eye towards this 'new' province of subsea intervention. Partially fuelled by the loss of the nuclear submarine 'Thresher' and the US Navy's enhanced interest in submarine rescue, they envisioned 'inner space' as a new frontier for the application of their awesome technical strength. Thus followed the introduction of numerous small manned submarines of various degrees of sophistication into an unknown and questionable market.

Although providing good observation characteristics and the advantage of horizontal range, the submersible's work capability was limited. More importantly, it was only the operators of such vehicles who began to recognize the magnitude of the problem of launch and retrieval of these somewhat delicate pieces of hardware. The submersible industry saw a brief upsurge in popularity with the introduction of the first commercial diver lockout type in 1970 and the utilization of specialized mother ships by Intersub and Vickers in the latter half of the 1970s, but they never succeeded in filling a significant niche in the offshore petroleum market.

There were many other technical developments in the sixties which were significant, but perhaps few were as important as refinements in lightweight gear. Whereas most of the public views the commercial diver as portrayed by John Wayne in heavy gear or through the eminently successful TV movies of

Cousteau and the use of scuba, the backbone of commercial diving is lightweight, hose-supplied equipment. Offering unlimited gas supply, hardwire communications, and secure tether to surface or bell, such equipment became the standard of the industry. Particularly notable in the period was the development of the 'Band Mask', employing the demand regulator, which was a significant advance over the early flowthrough, lightweight breathing equipment.

Refinement, Technical Extension and Application

The period of the seventies, while introducing several new technical advances, might be more clearly viewed as the time when the basic developments were first broadly applied to the market. This was particularly true in the North Sea, where greater water depth, low temperatures and adverse sea-states posed significant challenges to the subsea industry. Some areas, such as hyperbaric physiology, continued to provide new insights into unknown areas — saturation/excursion diving, deep excursion methods, intermittent oxygen therapy for decompression sickness, isobaric counterdiffusion, and hypothermia limits. The latter, which we now take for granted, was still an unknown area in the sixties. Today we routinely accept and comprehend what is required to assure adequate respiratory and body surface thermal balance, but it was not always so. Ultimately, as usual it was done the simple way — bathe the diver in surface-supplied hot water. More sophisticated techniques were tried including electrical and isotopic heat, but subsea operators are a practical group with a job to do and refinement and sophistication do not always track with commercial necessities.

The phenomenon of helium voice, vocal distortion due to the characteristics of helium, although attacked electronically in the 1960s, found practical electronic solutions in the 1970s. So too were strides made in underwater television (to prove what the diver was talking about), and the industry made significant improvements in repair techniques with the introduction of successful dry hyperbaric procedures for welding.

Push–pull equipment for breathing gas conservation finally began to achieve commercial fruition in the late seventies after initial, tentative developments in the mid-sixties. Similarly, a development of the twenties, the armoured diving dress, witnessed its contemporary revival using space age technology. More significant, however, was the impact on the industry caused by

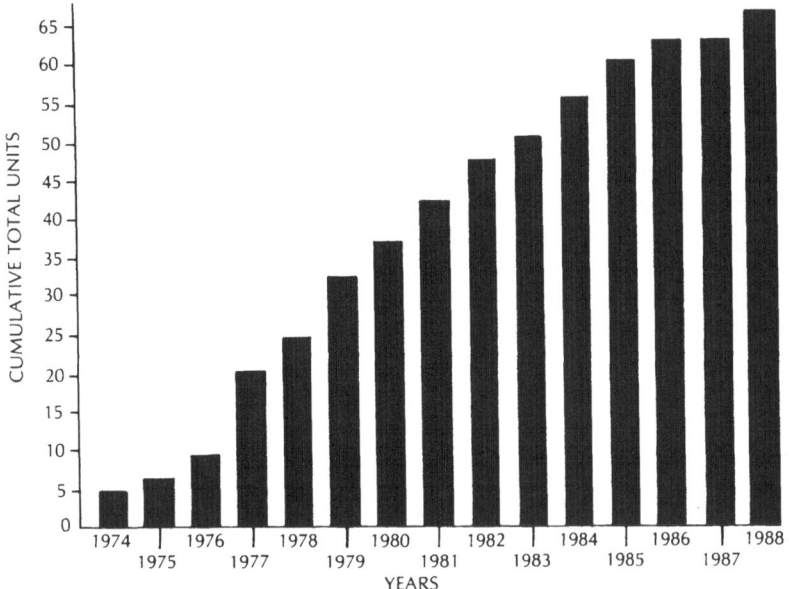

Fig. 1.4 Growth of the worldwide DSV fleet

the introduction of the DSV (diving support vessel) in the middle of the decade. The 'Artic Seal' with its dynamic positioning system and the 'Uncle John' with its semisubmersible stability revolutionized the nature of the platforms for subsea intervention. While still primarily a North Sea tool, these vessels and the fifty (Fig. 4) that have thus far followed have, to the amazement of some who have questioned their cost, broadly increased operational facilities — at a price which ultimately the customer has recognized as 'value-for-money'.

The Rise of the ROV

Few technical developments in subsea service have had as profound an impact on the business as the evolution and commercialization of the remotely operated vehicle (ROV). Although the ancestry of the current fleet of ROVs can be traced to the CURV series of USN vehicles, the first commercial ROV service began in the mid-seventies and employed the RCV-225. The 'flying eyeball', like most new developments, encountered some scepticism at first, and finding appropriate niches for its use took a few years.

Divers greeted the arrival of unmanned tethered submersibles with little enthusiasm since, particularly for visual inspection, they represented a perceived threat to their livelihood. Over a period of time, however, the industry has evolved a balance between the roles of the two — the ROV with unlimited underwater endurance for thorough visual inspection, the diver's dexterity for detailed examination and repair. The combination of the two capabilities has now become the norm for platform inspection; but pipeline inspection, once the province of the diver, is now dominated by the more cost efficient ROV.

The work vehicle, equipped with manipulators, took longer to find its place in the market. These larger and significantly more expensive ROVs entered service in the latter half of the seventies and found their initial roles in route surveys, bottom clearance/light salvage, and pipeline inspection requiring sophisticated instrumentation. During the early eighties they began to be applied to simple construction roles. Their real forte, however, has developed during the past few years in exploratory rig support. In this application their inherent advantages of lower cost, minimum space occupancy and weight, and small manning requirements have made major and accelerating inroads into a market formerly dominated by the use of deep diving systems and divers.

The technical evolution of ROVs has been extraordinarily rapid. In the span of less than ten years the industry has witnes-

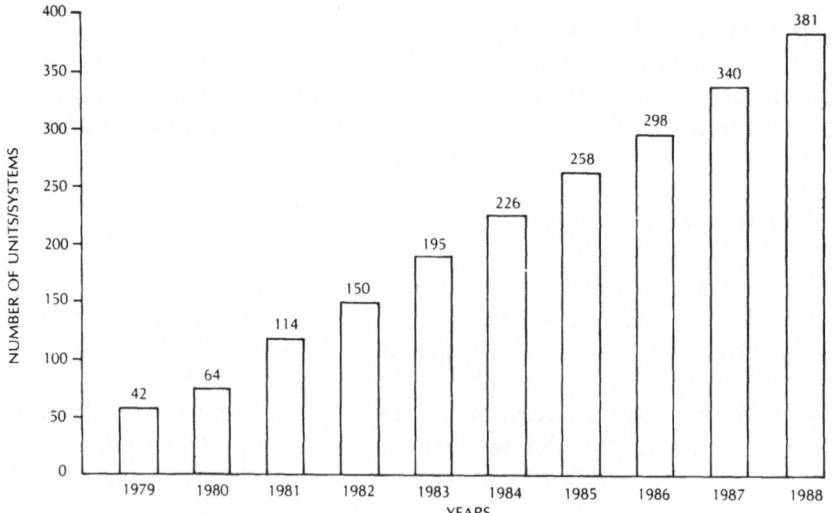

Fig. 1.5   Growth of the worldwide ROV fleet

sed the introduction of colour and low light level CCTV, precision locating and navigation systems, spacially correspondent and force feedback manipulators, fibre optic data transmission, cage development, and a myriad of refinements which have significantly improved reliability and utility. It is noteworthy, too, that the design and manufacture of ROVs is very much an international industry.

The number of ROVs in service (Fig. 5) has risen dramatically since their introduction. As new vehicle capabilities continue to evolve and the industry finds new applications, the number of ROVs can be fully expected to increase in the future.

SUBSEA PERSONNEL AND REGULATIONS

Perhaps some of the most striking changes which have occurred in the industry over the past twenty years concern the supply of diving personnel and the regulations governing their activities offshore. It is patently clear that we have not changed the intrinsic nature of the commercial diver; he is still an individualist with penchants for strong brew, fast cars, and similarly paced women. But how he enters the industry and the nature of his training have changed dramatically.

When the industry was founded, most commercial divers had been trained by the navy or had served an unstructured offshore apprenticeship learning on the job. Many did not have extensive knowledge of fundamental diving technology but, offsetting this, they had a wealth of experience in engine mechanics, piping, rigging, welding and other craft skills. Whether tender, diver or supervisor they were fully conversant with every piece of diving plant and quite capable of overhauling a diesel or compressor on site.

The need for qualified diving personnel began to outstrip the traditional sources of supply as the industry expanded. Consequently, today, commercial diving school graduates now dominate the labour force. The industry grew from an initial cadre of a few hundred divers to a current position in which there are over 4000 in the offshore industry throughout the world. With that growth has come increasing specialization. The industry now employs life support technicians, mechanics and chamber operators; no longer must the diver be able to hande all aspects of a diving installation. He is now better educated and often bell trained before beginning his commercial career, but because he

or she now generally comes from a background of amateur
scuba diving instead of manual skills, the diver who is the 'jack
of all trades' is rapidly disappearing from the scene.

It should be duly noted that government rule-making managed to keep pace with the expansion of the industry, and today
we have a host of regulations governing the conduct of subsea
operations. Although it is getting harder to remember the
period when there were no rules, at one point in time the industry was self-regulating. Insurance concerns and the spectre of
litigation under the American Jones Act and Longshoremen's &
Harbourworkers' Compensation Act provided companies with
their own motivation for the safe conduct of offshore operations.
However, the diving industry traditionally draws public attention disproportionate to its size and to be realistic, it is an inher-
ently risky business. Consequently, it is not too surprising that
today an international diving contractor must have procedures,
equipment and training standards which satisfy the OSHA,
USCG, NPD and DOE regulations established over the past ten
years.

THE BUSINESS TODAY

Status of the Industry

The subsea industry is slowly extracting itself from the depths
of the 1982/1983 recession. It is still ill, but the patient is no
longer in fear of dying. Margins continue to be depressed but
revenues are rising again. A few casualties, weakened by the
previous downturn, can be expected to result in bankruptcies
and further mergers.

Over the past two decades the industry has grown from a few
million dollars in annual sales to be a significant global activity
grossing over $700 million per year. The management of the
industry has matured with it and now reflects an established
sophistication in finance, marketing, operations and resource
control. These skills are essential in today's market with its
increasing customer demand for improved offshore cost and
time efficiency and continued expansion of fixed price contracting.

The industry has moved beyond the period of unbridled zeal
for new tools and unstudied diversification. Today the role of
various subsea intervention techniques has cleared significantly

and selection of the most appropriate tools, based upon both operational and economic considerations, has become more scientific. So, too, the errors of the past reflected in enthusiastic but ill considered diversification have not been forgotten, and gut feeling has been supplanted by studied reflection and concern for symbiosis.

The market sectors have shifted in their financial importance. Diving support of exploratory drilling, once the largest part of the market, was ultimately eclipsed by construction support. Both of these aspects of the business, however, have now been overtaken by the growing importance of the inspection, maintenance and repair (IMR) area. Over 4300 production platforms are now installed around the world, and an average of 300 are added annually. With a growing concern for their programmed inspection and maintenance by the oil companies, certifying and regulatory bodies, the IMR area (once the poor orphan of the business) has now emerged as the largest revenue producer.

Technology

The industry has made major strides over a relatively short time span in melding a wide array of scientific findings, technical developments and innovations to forge a solid underwater work capability. We have witnessed the gradual migration from pioneering efforts wherein each operation seemed new and untested to the evolution of routine, standard practice.

Of particular note has been the industry's early preoccupation with reaching ever greater depths. During this period it appeared that the need for greater operational depth capability was just around the corner. Many of these simulation dives provided significant scientific findings affecting the total spectrum of hyperbaric exposure; others merely established new depth records.

Ultimately it became apparent that the technical capability to place man at extreme depths did not equate with the real needs of the offshore petroleum industry. In point of fact the *average* depth of all commercial dives has increased only 6 feet per year over the past two decades. What we have come to realize is that considerations of cost, space, weight, manning requirements, logistics, training, job planning, proper tools, safety and alternative techniques are equally as important as the sheer technical capability to put a diver at a desired depth.

There are still numerous gaps in our knowledge of subsea

intervention, and many improvements in techniques and equipment are yet to be made. The difference today is that the industry is vastly more capable of separating scientifically significant avenues from the nice and the interesting.

The dissemination and sharing of technical information on a widespread, international basis has become a major goal of industry participants to assure the safety and well being of all. The early days of trade secret techniques have dissolved into today's spirit of co-operative activity. Reflective of this new era has been the recent, albeit abortive, collaborative effort of the 'Industry Initiative in Diving' wherein oil companies, diving contractors and others agreed to pool their resources to engage in a number of technical projects to raise the overall quality of diving operations. While the IDD might be perceived as a failure, it is best viewed as a harbinger of the pattern to come.

Image

In the early years of its development the industry had something less than a pristine image. Among the oil companies, divers were viewed as expensive, bubble blowing cowboys whose presence on a job was treated as a necessary evil and whose work in the invisible murky depths was of questionable quality. Some still think that way.

Over the course of time, however, the role of subsea intervention has taken its place as an important service in the offshore industry. Enhanced professionalism in the conduct of operations, equipment standards, and the training and performance of personnel have contributed significantly in elevating the industry's image to that of a mature, responsible business.

The AODC has played a major role in education, standards development, and coalescing the views of member companies to present a coherent voice for the industry. The UK DoE Safety Directorate has adopted a unique collaborative *vis-à-vis* adversarial role with the industry to develop a viable, realistic set of uniform operating requirements.

Regardless of the vast strides that have been made in establishing a solid image for the industry, it must be continuously borne in mind that because of the unique nature of diving in the public's mind, any accident results in a major outcry in the nation's press. Such incidents erode national confidence and tend to nullify all the effort which has gone into building a significantly improved record of safety.

CONCLUSION

The subsea business has emerged over the span of twenty years to become a significant industry. Yet, despite its growth rate and size, it still retains its entrepreneurial character, pioneering spirit, and aura of daring and adventure. Now if we could only figure out how to forecast and avoid those cycles in the E and P budgets of our customers . . . .

2

Concepts for the Future

H. G. Delauze, Comex Services S.A.

TRENDS FOR TOMORROW

In these times involving a decreasing market, shared by a very competitive industry, and over-supplied with diving vessels, it is time to take stock and look at the trends for tomorrow.

What are the signs on the wall?

- Narrower profit margins (for companies and contractors).
- Need to extend the work season in harsh environments.
- More technologically sophisticated interventions.
- Making human intervention more reliable.
- Extending the 'reach' inside structures.
- Deeper intervention.

Narrower Profit Margins

Over-supply of facilities and services in the industry, coupled with a weak market, is creating intense pressure on operating margins for even the most efficient of the underwater contractors.

Even those major diving companies who have re-invested substantial sums in new vessels and equipment are finding it almost impossible to earn adequate returns on these investments. The market wants the best equipment but it is less willing to pay the rates that the new equioment demands. An uncomfortable paradox.

So what is to be done? Obviously the drive for further increased efficiency must continue. More and more, the larger diving contractors with proven experience and the right kind of equipment will take work on a lump sum basis — guaranteeing clients that the task will be done within a fixed budget. Lump sum operations are a high risk business in financial terms but the major contractors must see this method of working as one of the few which offer a reasonable rate of return.

Extending the Season

Just as drilling rigs are now being built for all-weather, all-year round operations, in harsh environments such as the North Sea, so the underwater contracting industry has responded to the same concept of extending the work season. The new generation of fully DPd monohull DSVs such as Seabex and Seacom have demonstrated their ability to support complex construction operations in the North Sea winter (for example, Seabex 620 ft tie in at Magnus winter 82/83). Radical new ship designs, such as the ORELIA, are providing the monohull DSV with the stability and motion characteristics similar to that of a semi-submersible. Better designed moonpools, heave compensated bells, more sophisticated cursors and so forth are all contributing to more efficient, all-weather, longer-season performance. But it all costs money.

New Technology

The days when the most that was expected from a diver was to attach a shackle, cut a wire or dig the seabed with his bare hands in order to pass a sling under a pipe are, thankfully, gone.

Ever increasing sophistication is called for in both the planning and achievement of tasks. It has to be remembered that the function of diving is to put a qualified man into the subsea worksite where he can use his skills to the full. Thus NDT techniques, for example, require a lot more brains than brawn.

In ROV and robotic techniques we are advancing. But we are a long way from the day when the bulk of subsea operations will be handled by the so-called diverless systems. What we do need to do is to keep continuously improving our engineering preplanning, our training, our project briefings and familiarizations, and the reliability of our tools and equipment so that divers at the worksite can perform with maximum safety and efficiency.

Reliable Human Intervention

Considering the cost of diving intervention from a DSV is today around £4000 per hour, it is obvious that clients want every possible guarantee of the reliability of the operations to be performed subsea. To achieve this the underwater contractor must be able to deploy, in-house, a huge array of studies, procedures, training, quality assurance and control, equipment testing and engineering back-up.

Beyond the moral considerations, the cost of accidents in financial terms and loss of reputation alone is the best guarantee that serious contractors will not cut corners at any cost (despite tighter cost control and narrower profit margins). Operators and Government Agencies share this concern and the amount of paperwork involved today is positively amazing.

Expanding the Range of Operations 'Inside' Complex Structures

Given the sheer size of some production platforms in existence today and considering that very few of them do carry a permanent diving system onboard, diving interventions are normally performed from DSVs brought in for the purpose.

Due to the proximity of the structure, the presence of pipe underwater and the sheer amount of time involved, almost no DSVs today are operating at anchor. DP has developed to the point where its reliability can be trusted although there is no room whatsoever for complacency. It remains that in order to operate inside some structures while keeping a minimum safe distance between the vessel and the structure, and/or the diving bell and the structure, umbilicals of up to 70 metres and more have to be used. Factors such as the size of the bell, the autonomy afforded by the bail out bottle, and the size of the bail out bottle in relation to the size of the bell trunking are self-limiting factors.

Having stated that the maximum length of umbilical you can afford on a particular worksite is 76 metres, can you demonstrate to the client that 78 metres is impossible? Where do you draw the line?

Within Comex today, as much as is reasonably practicable,

this limit is established jointly between the Diving Super-intendent, the Project Manager, the Safety Officer and the Client and is, of course, subject to diver's acceptance.

Deeper Intervention

In the North Sea today, most dives are performed well within the 200 metres range. Some operations, not statistically signifi-cant, have been performed at greater depths. Yet it should be noted that complete welding operations have been performed at 300 metres and that some construction dives are taking place routinely off Brazil in this depth range.

Man is unlikely to work commercially at pressure in depths of thousands of metres — at least in the foreseeable future. Oil companies and seabed mining companies aiming to work at these great depths will, therefore, have to work with the underwater contractors to develop economic diverless interven-tion techniques. Before these techniques are refined to the point where they become economic at shallower depths, there will remain a role for cost effective diver intervention at depths above 1000 metres.

Operations down to the 400/450 metres mark can be performed today using standard heliox techniques. It must be acknowledged though, that this deeper penetration is accompanied by an ero-sion of the usefulness of the diver caused simply by the increased work of breathing necessiated by the increased density of heliox at those depths. In other words, the task of breathing occupies a greater percentage of divers' energy and brings about additional associated fatigue.

CONCEPTS FOR THE FUTURE

So what are the practical developments for man intervention for the immediate future?

I can see a *horizontal* one and a *vertical* one.

Vertical

This involves the capability to dive to greater depths (that is, below 300 metres) without the fatigue penalties associated with increased density of Heliox.

The way to reduce gas density is to go for a thinner, inert gas. The solution seems to be — Hydrogen.

We know what is to be gained by using this gas so what are the drawbacks of using it?

Explosiveness

This factor has, for many years, inhibited the development of hydrogen/oxygen mixtures for diving gas purposes. Extensive research performed in our hyperbaric centre at Comex has demonstrated beyond any doubt that below a critical level of 4% oxygen in hydrogen there is no danger whatever of combustion or explosion. This threshold is further reduced by the fact that, at the depths where we would consider using hydrox, we would never allow the concentration of oxygen to be above 2%.

Narcotic Effects

Hydrogen has a narcotic effect which appears to vary in its effect on divers. Experiments carried out by Comex in early 1984 have demonstrated that this problem can be mastered, whether through selection, adaption or use of trinary mixtures.

The experiments to which I refer were performed in our hyperbaric research centre in the main chamber of hydrosphere. The chamber was partly flooded and the experiment took place both in water and in a hydrox atmosphere in a bubble in the centre of the chamber, which facilitates observation.

Horizontal Plane

This will involve the use of a decent sized submarine allowing divers locking out from it to penetrate with relative ease, in terms of umbilical length, to the most remote areas of offshore structures. The submarine will have the endurance to travel to the worksite, get the job done and return to base without the support of a surface vessel.

This concept involves the resolution of a number of problems such as power autonomy, multiple system redundancy, energy storage for all normal life support functions plus emergency back-up and a reliable gas reclaim system.

Commander Cousteau had such a dream back in 1965. The pressure vessels for his 'Argyronete' were completed in 1970 when the project was stopped by the French Government for budgetary reasons. It must be noted that the state of the art at the time would not have provided this white elephant with enough power or autonomy to be of much practical use.

12 years later Comex decided to turn the dream into a hard reality thanks to the following practical technological advances:

- Energy generator capable of functioning underwater, the capability to store energy such as gas at a very high pressure or liquid oxygen storage.
- Use of composite materials so as to save weight whenever possible.
- Effective gas reclaim loop.

This futuristic animal, rebaptised the 'Saga', will have a practical range of intervention of 200 nautical miles and a useful intervention time of 15 days.

It will either carry a team of 12 personnel, 6 divers and 6 crew for diving intervention down to 200 metres for 10 days or 4 divers and 6 crew for a week down to 350 metres or 4 divers and 6 crew for 4–5 days to its maximum depth capability of 460 metres.

The development of this ambitious project is done in cooperation between Comex and CNEXO (The French National Centre for Exploitation of Oceans.)

The choice of propulsion is as follows:

- Closed circuit diesel engine.
- The Stirling engine (external combustion).
- Nuclear energy.

The 'Saga' will use 2 Stirling engines while the 'Saga 2' presently under design study for COGLA (Canadian Oil and Gas Lands Administration) will use a nuclear engine.

CONCLUSION

Praise is due to the efforts which over the past 10 years have brought the young diving industry into adulthood. We are currently operating in the 200 metres range and the array of platforms and pipelines out there in the North Sea bears witness to the contribution from the diver.

Prophets of doom tell us that having played a somewhat useful part in this achievement, the diver cannot be useful any further or deeper and must be replaced by machines.

I'll say this, divers are still going to be useful for a good few years to come and not only to retrieve machines and robots tangled up in their own lines!

The human brain, eye and touch still has not been duplicated satisfactorily. Divers are better educated, qualified and more useful than ever before and the technology is there to take them further and deeper safely.

3

The Future Role of the Diver

Keith T. Bentley, Phillips Petroleum Co.
Europe–Africa

INTRODUCTION

As a result of the last few years having witnessed the rapid growth in popularity of a variety of underwater vehicles, it might well appear to the uninitiated that there may be a day, not far distant from now, when the diver might be replaced completely.

It is my view that those more closely involved have reached something of a dichotomy of opinion regarding the future role of the diver. This situation has been created by the rapid increase in subsea requirements of many disciplines taking place simultaneously and, in a large number of cases, independently of each other. This state of affairs has been catalysed by the ferocity of competition in the industry, which has been continuously on the increase in recent years and which in itself has contributed to the reduction of cross-fertilization of ideas and philosophies between contractors and clients.

As we find ourselves amid a buoyant offshore European market in a reasonably settled state, it is perhaps opportune, more so than at any previous time, to appraise the diver's services in the coming years and compare his role in relation to a possible machine alternative.

DIVING OPERATIONS

The diver is still the most dextrous mode of intervention in the oceans. However, as we are all aware, there is a depth limitation that will be arrived at beyond which he will be unable to operate, and prior to that a shallower limitation beyond which point his safety and efficiency will be brought into question.

Current views as to these maxima are postulated at being between 360 and 460 metres (1200–1500 feet). More realistically, I think we are currently looking at something between 300 and 360 metres (1000–1200 feet) when we simultaneously take into account the safety and cost factors involved at these depths. From whichever point of view, a definite plateau of limitations arrives on exceeding 300 metres.

As many are aware, we have sent divers to record depths in the region of 686 metres (2250 feet) under hyperbaric conditions, which gives us the possibility of commercially reaching these depths in the future. But remember that the transition from laboratory conditions to commercial application in open water at these depths is something of a 'quantum jump'.

As is often the case with commercial enterprise these days, the question of whether or not a particular mode of subsea intervention is or is not utilized is influenced by the cheapness of the service offered.

Because of the nature of the diving operations in the underwater business, the deeper the diver goes the more it costs. This, coupled with the physiological limitations imposed on him, which again increase with depth, and the hazardous medium in which he works, results in many clients undeavouring to remove him from the water at the first opportunity and to replace him with some form of automatic or remote intervention. It is also fair to say that the diver's vulnerability does not enhance his competitiveness when placed alongside the ever-improving alternatives offered.

During the first half of this decade we have found ourselves arriving at a 'watershed', in that amongst other things the following developments have occurred:

(i) Exploration and production have exploited deeper waters than the previous decade's average maximum of 180 metres (600 feet).

(ii) Remote subsea completions, floating production systems

and automatic wellhead concepts have become serious con-
tenders to conventional platform production techniques.
(iii) Vehicle intervention has been consolidated in the eyes of
the commercial oil industry and has clearly established a
permament slot in the subsea market.

As a result of these developments, and a large variety of special
projects conducted within the oil companies, suggestions for
specialist intervention via all modes have been presented.
Wherever possible there has been an attempt to capitalize on
providing a diver alternative.

However, throughout these shifts towards new approaches,
the diver has continued to maintain his presence at a more-or-
less even pace and, in some notable cases, come to the rescue of
failed alternative interventions.

NON-DIVER INTERVENTION

Both remotely operated vehicles (ROVs) and one-atmosphere
systems (manipulator/observation bells, atmospheric diving
suits (ADS) and manned submersibles) are low-dexterity
machines.

The most successful and established market for unmanned
intervention at this time is that of drilling support, where we
have witnessed during the early 1980s a take-over of the role by
both ROV and ADS type vehicles. Requirements of intervention for drilling support are limited (in the main) to an established finite number of functions. Typical tasks are bull's-eye
debris clearance, guide wire stabbing or removal/replacement
and seabed surveillance. Such types of tasks, being of a low-
dexterity nature are well suited to these vehicles and no doubt
their use will continue to increase in momentum as the years
progress.

The larger vehicles have proved themselves as good pipeline-
tracking survey tools and because of their endurance on the sea
floor and the ability to carry a complete package of visual,
acoustic and electronic surveillance paraphernalia, have become
increasingly popular. Again however, low dexterity, if not zero
dexterity intervention is provided.

Beyond these two spheres of work (drilling support and
pipeline survey) the encroachment of ROVs into the diver market has been successful, but severly limited in application. Suc-

cesses of note have been those of the small 'eyeball' vehicles performing visual structural survey roles and providing *support* to the diver, acting as an excellent visual intermediary between the surface and the work-site.

Some vehicles have performed low-dexterity cathodic protection functions, and some intermediate-to-large vehicles have performed jetting and brush cleaning operations in and around fixed platforms. However, any work that has needed a degree of dexterity akin to that of a diver has not been successful.

DISCUSSION

Rather than a complete 'take-over' situation developing within the existing commercial depths encountered by oil industry, I see the vehicle market working in harmony with the diver as we are witnessing with the vehicles' use in observation and cleaning roles.

The 'watershed' and the three prime contributory factors mentioned earlier, and the resultant varieties of specialization within oil company project groups, have resulted, in my view, in the dichotomy of opinion as to what should be used, and how, and to what depth. This was probably inevitable, and as the dust settles and the options available will take their respective courses, as other evolutions have done before within the industry.

Returning therefore to the diver and his unique and unsurpassed quality of dexterity, it is clear that certainly within the air range of operations to 50 metres (165 feet) he is never likely to be seriously challenged.

Within this depth range he is an extremely reliable, cost-effective and efficient form of intervention. Ironically, however, he is also within the depth range that has produced the majority of accidents in recent years. Working in a much more hostile environment than his deep-sea counterpart, generally without a diving bell refuge, and usually in isolation with, more often than not, zero visibility, has made him more vulnerable to these accidents. As we are aware these accidents have not enhanced the client's view of the diver's credibility.

From 50 metres down to the commercial maxima of around 180 metres currently encountered in European, US and Middle Eastern waters, I believe the diver will, despite the inroads made by the vehicle market, maintain a steady percentage increase per annum over the next few years. The continued increase in requirements for structural repair and maintenance will ensure the long-term requirements of high-dexterity intervention which only the diver is able to provide.

At the deeper end of the diver spectrum, it has been interesting to follow the hyperbaric welding programmes that have been conducted in Norway at depths of around 350 metres during the last two years. These trials have undoubtedly consolidated the merits of diving intervention in being, first, able to reach such depths, and, second, able to perform satisfactorily at those depths. The knowledge gained from these operations will I am sure provide a solid base for the future of the working diver at the bottom end of his depth limitation.

Safety

At this point, I would like to touch on the question of diver safety and the industry's attitude to this problem.

It is inevitable, as with any major airline disaster, that any diving fatality is a prime target for the media. Thus it would not take too many accidents before the question of diver safety was seriously contended. It must be recognized, however, that those within the commercial diving world operate in a relatively high-risk area where a certain amount of risk has to be accepted. The application of safe practice and vigilance thereof must go hand in hand to keep trouble to a minimum. I witness much lip-service given to safety within diving operations both from operators and clients, and the safety aspect of diving being used as a convenient scapegoat for the use of other forms of intervention. The fact is that the safety record of commercial diving operations is extremely high, indeed better than that of the aviation world — whose own standard of safety is already high in itself. It is just a sad fact that when things do go badly wrong, divers, like airline passengers, have little margin to play with, and are usually in an all-or-nothing situation regarding their survival.

The author Nevil Shute noted in one of his books published in the early 1930s that experience had taught him one particularly sad fact — that you cannot sell safety. This statement is as true today as it was then. However, there is no doubt that conscientious effort has indeed been made to establish clear standards of safety, and to ensure continued monitoring of their upkeep, and improvement.

CONCLUSION

It is very easy for the oil industry to take the diver for granted. He works within a large and fragmented worksite covering all corners of the globe. He is rarely seen at work and perhaps, therefore, is not given the credit that he deserves. He is by nature an individualist and something of an extrovert. Indeed, he may well have to be of this nature to sustain tolerance of the intermittent and varied geographical locations and physiological pressures into which the industry thrusts him.

Let it be clearly noted that without him, all that we have witnessed offshore during the last twenty years of exploration and subsequent production would not have come to pass. Furthermore, if there were any subsea aspect that could have succeeded without him, it would no doubt have taken many times longer at many times the price.

I am confident that in a paper in twenty years from now on this same subject, I would conclude with similar words.

4

Programme for Working Dives to 400 msw
I. Aspects of the Human Factors

Y. Giran, MD, Comex Services, Marseille, France

INTRODUCTION

Despite the large amount of experimental data obtained from exposures of human subjects to simulated chamber dives in excess of 400 msw, only six men really dived in open sea, in that depth range (Janus IV, 460 msw open sea dive Oct. 1977). An excursion dive even took place down to 501 msw during the same diving operation.

Until now more than 40 divers have been exposed to depths of 400 msw or deeper during the French deep diving experiments.

During the Atlantis III dive at Duke University, three subjects were successfully compressed down to 686 msw.

In 1983, the Deep Dive Development Project demonstrated that both diving and hyperbaric welding tasks can safely and efficiently take place on a routine basis at 300 msw.

Safe working dive to 400 msw is considered physiologically feasible provided that technical and medico-physiological aspects are carefully evaulated, controlled and tested in an on-shore trial dive.

We shall discuss the main topics to be taken into account in the definition of the human factors programme for working dives to 400 msw.

DIVERS SELECTION

This essential aspect is mainly based on professional and medico/physiological criteria as well as divers' training.

The professional criterion chiefly covers an extensive professional and technical experience in the various tasks intended to be done at depth. Moreover, it also includes an acceptable sociable attitude which is essential to create a team spirit.

On the medico/physiological side, despite various trials, it has not yet been possible to establish sure selection criteria for very deep divers. Nevertheless apart from the standard medical examination for divers the selection will essentially consider the following topics:

- Central Nervous System (Clinical Neurological Examination, Neuropsychological evaluation, Electroencephalography)
- Lung function evaluation
- Cardiovascular adaptation to muscular work (Maximum Oxygen Uptake)
- Cochleo-vestibular system evaluation (Electronystagmogramme and audiometry).

A minimum level of training appears to be necessary for deep diving. Apart from the technical training, it should include:

- Some physical training (ventilatory work against breathing resistances)
- Emergency medical training for divers
- Exposure to pressure in an onshore controlled environment.

CENTRAL NERVOUS SYSTEM

Different researchers have reported some potential minor and reversible changes in the Central Nervous System both during and after deep diving. This leads to careful evaluation of the Central Nervous System reactions by means of standardized batteries of tests.

During the compression phase to depths of 400 msw, an adequate compression profile enables minimization of the so-called 'High Pressure Nervous Syndrome' which is to be carefully controlled on the other hand during an onshore trial dive by means of neurophysiological investigation techniques.

In the post dive period, a complete CNS follow-up programme is expected to provide valuable information on the neurological short term effects.

RESPIRATORY SYSTEM

As gas density increases with depth, it leads to a modification of the gas flow in the bronchial tree (dynamic resistances). As a consequence, the maximum expiratory outputs are decreased and the work of breathing increased to maintain adequate outputs necessary for alveolar ventilation, particularly during muscular exercise.

In other words, the muscular work capacity of the diver decreases when depth increases, due to a ventilatory limitation. On the other hand, it appears that a certain CO_2 retention level is not unusual in divers at great depth.

It has been clearly established that adequate ventilatory training against resistance at surface level improves the diver's capacity to maintain an acceptable CO_2 level and to cope with the ventilatory limitation. Such training appears to be beneficial for working dives to 400 msw.

THERMAL BALANCE

Thermal exchanges are largely increased in deep saturation diving. They take place at the skin level which is about 1.8 m^2 as well as at the respiratory tree level which represents a surface of exchange of about 100 m^2.

The thermal comfort sensation which is provided by a thermic balance in equilibrium essentially depends on the physical properties of the medium (inert gas conductibility, absolute pressure, water temperature, gas and water flow).

Exposure to water increases by about 50% the thermal exchange compared to exposure to air.

Moreover, with respect to the respiratory exchanges, the combined effects of the high thermal conductibility of helium and of the High Gas Density in deep diving have led the researchers to describe the so-called 'Respiratory Heat Haemorrhage'.

For the same gas, it appears that respiratory heat exchanges are proportional to absolute pressure and ventilation.

Furthermore, as thermal comfort is basically subjective and variable and because of the potential risk of uncontrolled hypo- and hyperthermia, adequate thermal monitoring at diver's level is considered essential in deep diving operation.

HUMAN FACTORS ASPECTS IN WELDING HABITAT

Apart from the usual aspects related to dry exposure to depth the welding activity introduces several other points of concern.

An acceptable breathable atmosphere can be obtained when MMA and MIG are used with respect to carbon dioxide, carbon monoxide, nitrous oxides, fumes, particles and heat, provided that these pollutants are extracted and that careful continuous monitoring is performed.

The use of the TIG welding process leads to introduction of argon and production of ozone within the habitat atmosphere. Potential risks are acute intoxication (ozone), narcosis and inert gas counter diffusion effects (argon). Currently, the use of a full face welding mask is therefore necessary when welding on TIG.

SHORT AND LONG TERM EFFECTS ASSESSMENT

It is chiefly based on the statistical analysis of an intensive medico/physiological programme repeated several times after the dive (approximately one week, four to six weeks, six months and one year). This programme basically includes:

- Central Nervous System evaluation (EEG, clinical examination, neuropsychological investigation)
- Lung function tests and cardiorespiratory adaptation to effort test
- Cochleo-vestibular system evaluation
- Biology
- Standard medical examination for divers.

CONCLUSION

As fas as physiology is concerned, human diving to 400 msw has been shown to be feasible since the seventies by means of simu- lated chamber dives. Many research programmes have improved our knowledge on the physiological adaptation of human beings to great depths and shown the feasibility of safe diving practice at 400 msw.

Nevertheless, it is clear that for working dives to 400 msw a careful evaluation should be done of the medico/physiological aspects to gain more data and assess both short and long terms effects.

On the other hand, experimental data obtained from recent hydrogen simulated chamber dives indicate hydrogen as a potential alternative inert gas for future deep diving.

Programme for Working Dives to 400 msw
II. Technical Aspects

J. P. Imbert Comex Services, Marseille

SUMMARY

In order to increase its depth capability for offshore operation to 400 msw, Comex has initiated a Deep Diving Programme. One of the development projects concerns the diver's individual equipment. To provide the diver with safe and efficient equipment, the project proceeds through definition of basic requirements, specifications of performance standards, selection/development of equipment, unmanned tests, offshore tests at conventional depth and a final deep onshore dive. Standards have been defined according to Comex experience, for breathing performances, inspired gas temperature, inspired CO_2 level, reliability, gas consumption and efficiency. Though the project is still in progress, some results are presented that concern the qualification of the Comex gas reclaim system.

NOMENCLATURE

Standards: The word standard is used throughout the text to represent the performance goals and tests procedures used for equipment evaluation.

HPNS: High Pressure Nervous Syndrome.

UBA: Underwater Breathing Apparatus.

BIBS: Built In Breathing System.

DP/V loop: x/y record of respiratory pressure versus tidal volume ventilation used during breathing performance studies.

RMV: Respiratory Minute Volume.

INTRODUCTION

Conventional and Deep Diving

At this time, one can consider that conventional diving does not exceed 200 msw. It represents most of the diving activity and is typical of North Sea diving.

TABLE 1
Comex saturation diving activity in North Sea [a]

	Number of saturated divers	Time in saturation (h)	Number of bell runs	Time at bottom (h)
1983	656	272 000	5 459	34 880
1982	380	121 000	2 492	16 620
1981	447	163 000	3 232	22 100
1980	373	150 000	2 777	19 270
1979	403	128 000	2 930	18 200

[a]Units: time in men × hours

Operations from 200 to 300 msw are today considered as deep diving. One of the early deep dives was performed by Comex divers who recovered a well head at 326 metres in Labrador in 1975. However, activity in that depth range has only started on a routine basis since 1983.

TABLE 2
Comex deep diving activity in Brazil after one year of operation in the Campos field [a]

Diving depth	Number of saturated divers	Time in saturation (h)	Number of bell runs	Time at bottom (h)
200–250 m	24	14 850	69	408
251–307 m	43	24 030	107	554

[a]Units: time in men × hours

300 msw diving represents the state of the art and its importance will certainly increase in the years to come with the development of projects on the edge of the Brazilian continental

shelf and in the Norwegian trench. Considering the next step, which is to propose diving services to 400 msw, we must first admit that:

- Our offshore experience reduces to 10 hours working time performed by the Janus IV divers at 460 msw in 1977.[1]
- Each additional depth increment brings along its own specific problems. It seems that most of the technical solutions developed recently for 300 msw will fail to match the 400 msw specifications.
- Though diving methods tend to minimize fatigue and risks, divers at 400 msw will be exposed to HPNS, thermal stress, long saturation time and increased gas density and will still be asked to carry out almost the same tasks as in conventional diving.

Therefore, we think that the standards prepared for 400 msw diving must be kept very high.

Comex Deep Diving Programme

In order to increase its depth capability for offshore intervention to 400 msw, Comex has initiated a Deep Diving Programme.

The programme is designed for the development of suitable life support equipment, operation procedures and work techniques, as well as personnel selection and training.

However, the aim of this presentation is to focus on one aspect of this programme, which is the diver's individual equipment project.

The objective of this project is to bring about the systematic development and/or validation of equipment necessary for safe and effective underwater operations.

Organization of the project is presented below along with some of the early results.

METHOD

To demonstrate work capability to 400 msw, the project is organized in different steps that allow progressive assessment of the problems and solutions:

1. Definition of basic requirements.
2. Specification of minimum acceptance criteria and of recommended performance standards.
3. Review of existing equipment capabilities.
4. Short list of equipment that meets or has the potential of meeting the performance standards.
5. Alternatively, development of new equipment.
6. Unmanned testing of equipment in an onshore facility.
7. Manned testing on selected worksites operating at conventional depths.
8. Manned testing in an onshore facility of equipment, personnel and procedures to 400 msw.
9. Follow-up of deep operations in order to gain experience and eventually improve procedures or equipment.

At any moment, new findings may arise that may require modifications of the equipment and reset the programme to an earlier step. This very classic procedure allows a complete definition of the equipment through:

- Technical performances.
- Reliability figures.
- Diver acceptance.
- Efficiency (gas reclaim).
- Operational cost and maintenance.
- Track records and safety records.
- Operating manuals.

Definition of Basic Requirements

A diver's individual equipment must be capable of supporting him in normal safe conditions and must allow him to cope with emergency conditions.

Normal Working Conditions

Environmental conditions are 400 msw, 4 to 7°C ambient water temperature. Latest saturation procedures designed for 400 msw according to Comex deep diving experience propose the following guidelines for divers' intervention:

- 3 hours lock-out time in water.
- one dive a day.

- 3 divers' teams (one day bellman, two days locked-out).
- 10 to 14 days at bottom (to be defined according to regulations).

Breathing gas mixture is heliox. Because a diver's actual gas consumption would be around 100 m³/hour, a gas reclaim technique is to be used.

Risk Analysis and Emergency Procedures

The basic statement is that there must be a way to monitor diver, equipment and environment to control that the diving conditions are safe. In case of any deviation, specific equipment/procedure/personnel should allow the safe recovery of the diver into the bell, which is the basic emergency procedure for saturation diving.

Safety analysis of the deep diving situation does not differ much from the conventional diving operation, except that:

- potential HPNS may repair a diver's judgement or capabilities to react,
- distance from surface makes controls and assistance difficult.

Table 3 lists the main emergency situations that a diver might have to encounter. Because our topic is individual equipment, it only considers the diver's side of the problem. It is obvious that a loss of breathing gas supply, for instance, would also be handled at bell and surface level which proposes all sort of control, alternative sources or back-up systems.

Diver Monitoring

Diver monitoring during offshore operations has several limits.

The first one is the diver himself, in the sense that he would not stand uncomfortable electrodes, nor admit any cumbersome addition to his equipment. For that reason diver monitoring is restricted to safety control and, in most cases, excludes any scientific data collection.

The second difficulty is that some factor such as HPNS cannot be easily quantified. As a rule of operation, it will be assumed that HPNS effects have been thoroughly studied and calibrated during experimental dives and that they are kept to a minimum as long as the procedures are followed.

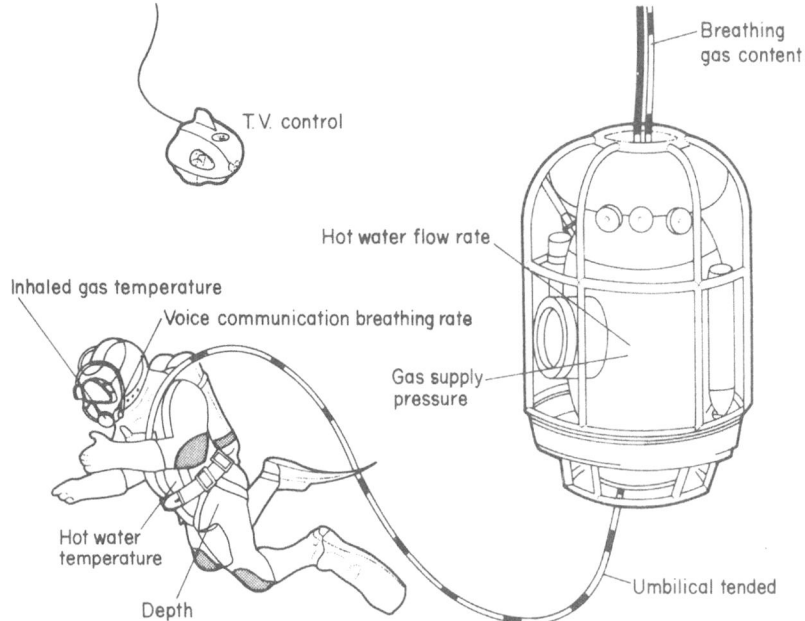

Fig. 4.1   Schematic diagram of diver monitoring

As a third limit, we have to consider that in some cases there is a gap between what we would like to monitor and what present technology can achieve. For instance, the key parameter, for the evaluation of the diver's thermal state, is the core temperature. Inhaled gas temperature is also important since respiratory losses represent a major part of man heat losses at 400 msw. Inhaled gas temperature can be readily measured by sticking a thermal probe in the diver oronasal but non-invasive core temperature measurement does not seem to be operational with today's technology. If no practical answer is provided by the time 400 metres diving occurs, we will have to accept to indirectly monitor the diver's thermal state by hot water supply control. Several experimental dives have permitted us to define the rule of operation of a diving suit/gas heater, and their results will be used to convert safe core and skin temperature in terms of hot water flow and temperature to the diver.[2,3]

A last limit concerns the level of qualification of the surface team. It is quite impractical to expect that the team supervising routine operation should include expertise which allows analysis of complex physiological parameters. For example, there is no doubt that cardiac rate would be an accurate way of monitoring

TABLE 3
Diving risk analysis for a working diver at 400 msw

Risks	Safety controls	Emergency procedures
Loss of TV control		dive aborted
Loss of communication	permanent voice contact	dive aborted
Loss of gas supply	supply pressure at bell and at surface level	diver on bail out bellman intervention emergency gas supply line
Gas reclaim failure	surface control of gas reclaim	diver on open circuit
Wrong gas supply	breathing gas analysis	diver on bail out bellman intervention
Loss of hot water supply	hot water temperature at diver level	dive aborted bellman/partner intervention
Hypothermia or hyperthermia	hot water temperature at diver level inspired gas temperature hot water flow hot water flow rate to diver	dive aborted bellman intervention
Entangled umbilical	ROV monitoring umbilical tended by bellman	diver on bail out bellman/partner intervention
Exhaustion	breathing rate via communications	dive aborted
Injury, illness Unconscious diver	communications ROV monitoring	dive aborted bellman/partner intervention

divers' effort and fatigue, but short term variations in heart rate may be most misleading and depend on a variety of stresses.[4] On the other hand, years of experience have trained diving supervisors to evaluate divers' workload via the breathing sound pattern that comes off the ear phones. Good and reliable communications are regarded as the prime requirement for diver physical effort monitoring.

Taking into account the above considerations, the minimum requirements for diver monitoring reduce to:

- Permanent TV monitoring by ROV.
- Voice communications.
- Breathing rate (via communications).
- Continuous breathing gas analysis for oxygen and CO_2 content (at surface).
- Breathing gas supply pressure (surface and bell).
- Hot water flow to diver.
- Hot water temperature at diver level.
- Inhaled gas temperature.
- Distance from bell (tended umbilical).
- Diver's depth.

Core temperature and heart rate monitoring would only be regarded as recommended.

Diver Emergency Equipment

Open circuit. This system should allow safe return of the diver to the bell in case of failure of the breathing equipment to function in gas reclaim mode. Open circuit diving should also be considered as an alternative to gas reclaim for continuous operation after a gas reclaim break-down.

Bail out system. The system should provide breathing gas at a safe temperature, sufficient to support the diver until the bellman can reach him in an emergency. The bellman should be equipped with an extra gas line integrated in his umbilical and specially designed to be connected to the bail out system of the diver in order to allow him to return to the bell.

Other auxiliary diving harness or suit should present a lifting ring which permits the diver to be hauled or winched into the bell when unconscious. The diver should wear a knife or a cutting device which allows him to disengage from a snag or cut his umbilical to free himself.

Standards

Performance criteria are considered as a goal to be used for design purposes. In case equipment fails to meet the criteria, its acceptance remains the responsibility of the Diving Contractor and Operator.

Breathing Performance Standards

There has been a considerable amount of literature published on underwater breathing apparatus performance goals and test procedures since 1980. Several schools of thinking have arisen and unfortunately there is no clear international consensus on what the performance standards should be.

Present Comex standards are based on the US Navy publication[5] and the conclusion of the Chairmen of the 1980 Bergen workshop on individual equipment.[6]

Our contributions came from experience gained after several years of intensive tests of individual equipment and were introduced in order to simplify the data analysis.

The first point concerns the classification of the UBA into two categories according to their operational use:

- *Category 1*. Equipment used for diving work, or diving emergency. It includes diver's gas reclaim and open circuit, bellman's mask and diver's bail out system. Equipment is tested with high simulated ventilations (40 to 75 l/min).
- *Category 2*. Equipment used for light activity. It includes BIBS, overboard dump system for BIBS, and welder masks. Equipment is tested with low simulated ventilation (22.5 to 62.5 l/min).

The second point concerns the calculation of the work of breathing. As the scope of the proposed Comex standards is restricted to the demand regulator type of equipment, the definition of the work of breathing is not as critical as for semi-closed or closed circuit. We therefore decided to define its calculation simply 'as the area of the DP/V loop', regardless of the reference line position.

A more elaborate definition can be used in a discussion but this will be left to the author's responsibility as no international agreement exists on it.

A third point is related to the maximum breathing resistance which is set to 1.4 kPa independently to the simulated ventilation. Our experience is that the maximum breathing resistance is critical for the higher ventilation rate, but is irrelevant for the lower ones, where the work of breathing becomes the principal criterion. This simplification is consistent with US Navy practice|for the demand regulator type of equipment.[5]

A last point, of less importance, is that the 75 l/min RMV is obtained using a 3 l tidal volume amplitude in order to separate the results on the DP/V loop.

The detailed Comex Standards for breathing performances are presented in Appendix I.

Inspired Gas Temperature Standards

These specifications are designed for gas heater evaluation. They define minimum and comfort temperature of inspired gas for a diver operating with an hot water suit. Data provided are based on a NUTEC publication,[2] J. Morrison's study for DOE[7] and some unpublished Comex data.

The detailed Comex Standards for inspired gas temperature are presented in Appendix I.

Core Temperature Standards

Diver's core temperature variation should be less than 0.5°C for the duration of the dive.

Inspired CO_2 Standards

Refer to Appendix I.

Reliability Standards

Reliability can be evaluated through offshore experience for existing equipment but needs to be demonstrated through specific unmanned test for new equipment.

Comex standards for a reliability study are presented in Appendix II and correspond to the Code of Practice of the American industry.

Breathing Gas Consumption Standards

A study was performed over several Comex worksites operating in the North Sea, in order to determine the average diver's gas consumption using conventional open circuit helmet in saturation diving. Data summarized over 500 hours of bottom time. The average diver consumption evaluated over the dive time is displayed in Fig. 2 in relation to the number of dives monitored.

The mean value is 30 l/min (or 1.8 m³/hour) at bottom pres-

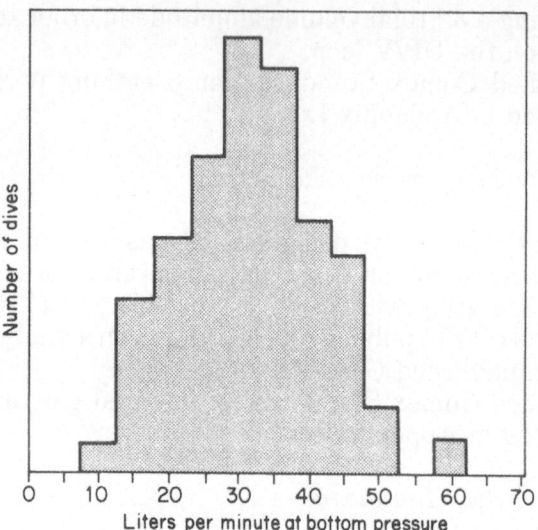

Fig. 4.2   Diver's average gas consumption

sure. Extreme values to the right-hand side correspond to incorrect equipment use or failure.

The mean diver's gas consumption appears lower than the classic 1.5 feet³/min (42.5 l/min) which thus seems to correspond to a round up figure. Comex policy is to consider 30 l/min as a standard diver gas consumption for routine operations and 40 l/min as an upper value to be used for emergency situations.

Bail Out System Autonomy Standards

The diver's bail out system is required to provide sufficient gas autonomy and inhaled gas temperature to support the diver until the bellman can reach him.

The minimum bail out autonomy is set to 3 minutes using a 30-metres long diver's umbilical. The associated breathing gas volume should be calculated using a 40 l/min gas consumption. The inhaled gas temperature should exceed the minimum safe inhaled gas temperature defined in Appendix I.

Gas Reclaim Efficiency Standards

Efficiency expressed in terms of percentage of gas saving is only relevant for the gas reclaim system. A survey of Comex worksites operating a gas reclaim system, mainly the Comex and the

Krasberg system, has shown that the average diver gas consumption ranges from 2 to 10 l/min depending on the diver's experience and dive conditions. Performance is about the same for the different systems used, because the sources of gas losses are identical: filter purges, leak on helmet neck seal, or switch to open circuit when returning to the bell.

Emphasis is put on the reference value used to calculate the gas saving ratio, because a 5 l/min actual gas consumption could lead to 66 or 88% gas saving ratio depending of the standard consumption considered. Taking a 30 l/min gas consumption as a reference, today's state of the art in gas reclaim seems to be 75% of saving.

Results

Programme is still in progress but some good results have already been obtained with the Comex Gas Reclaim System. After a series of tests and modifications performed in 1983 on a system specially designed for deep diving, positive results were finally achieved:

- One year of operation onboard Ugland Comex I for conventional diving.
- Unmanned test on breathing machine. Data obtained show that the equipment meets the requirements for breathing performances and inspired CO_2 level up to 450 msw. See Fig. 3.
- Associated with an integrated gas heater, a thermal study showed that the system succeeds in matching the comfort temperature for inhaled gas.
- Reliability test. No failure or warning was recorded during or after 240 hours test at 450 msw. See Fig. 4.
- The system is presently in operation onboard RIG 13 in Brazil, for dives ranging from 280 to 307 msw.

CONCLUSIONS

An effort has been made in the definition of standards for the evaluation of equipment for deep diving. Obviously these standards will be modified as experience is gained and further specific physiological studies are performed. However, these standards are, and will remain. very high and a lot of development and test would be required to provide the diver with acceptable equipment for 400 msw operations.

REFERENCES

1. 'Janus IV en mer', *Comex Internal Report*, 1977.
2. A. Pasche, J. Onarheim, S. Gordon, H. Padbury and B. Holland, 'The Comex onshore trial dive to 350 msw at NUTEC 1983: Diver heating'. *NUTEC Report* No. 27.83, 1983.
3. A. Pasche, S. Tonjum, C. Olsen and B. Holland, Deep ex 81: Project IV, 'Thermal study. Subproject:Diver heating'. *NUTEC Report* No. 8.82, 1982.
4. R. R. Pearson, 'Why do we need diver monitoring?'. *Proceedings of the International Conference Divetec'81*. Society for Underwater Technology, Workshop B, 24–26 November 1981, London.
5. J. R. Middleton, and E. D. Thalman, 'Standardized NEDU unmanned UBA Test procedures and performances goals'. Navy Experimental Diving Unit. Report No. 3.81, 1981.
6. B. H. Hjertager and T. Nome, 'Conclusion for the Chairmen'. In: Workshop on Diver's Breathing Equipment, Bergen, 30 November–1 December.
7. J. B. Morrison, 'Draft guidance notes on minimum performance requirements and standard unmanned test procedures for underwater breathing apparatus'. Draft guidelines of Department of Energy, 1983.
8. C. A. Piantadosi, 'Respiratory heat loss limits in helium oxygen saturation diving'. Navy Experimental Diving Unit, Report No. 10.80, 1980.

Appendix I

BREATHING PERFORMANCE STANDARDS

Scope of Standards

Standards are designed for demand regulator types of equipment.

Definition of Terms

UBA Classification
UBA are classified into two groups according to their operational uses.

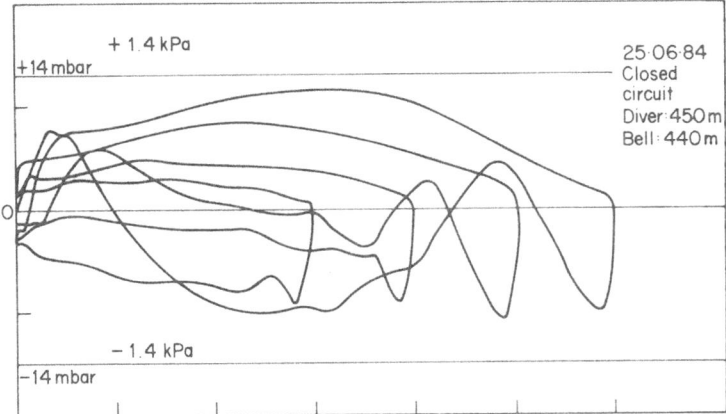

Fig. 4.3 Breathing performance study of Comex gas reclaim system according to test procedures described in Appendix I. Results presented were obtained at 450 m diver simulated depth and 440 m bell simulated depth. Diagram displays respiratory pressure versus tidal volume for ventilation rates ranging from 22.5 to 75.1 l/min. Upper part of the loop corresponds to exhalation, lower part to inhalation. The maximum respiratory pressures recorded and work of breathing computed were within the specifications

Category 1: Test ventilation from 40 l/min to 75 l/min RMV. It includes: diver's gas reclaim system, diver's open circuit system, bellman mask.

Category 2: Bell ventilation from 22.5 l/min to 62.5 l/min RMV. It includes: BIBS, overboard dump system for BIBS, and welder masks.

Work of Breathing
The work of breathing is defined as the area of the DP/V loop, regardless of the position of the reference pressure axis. Measured in Joule/litre.

Respiratory Pressure
(DP) The respiratory pressure is the differential pressure measured at the diver's mouth, during inhalation and exhalation, relative to the system reference pressure. Measured in kilo pascal.

Reference Pressure
The reference pressure is the pressure on the delivery side of the demand valve when the respiratory muscles are relaxed (i.e.

no workload) and therre is no respiratory gas flow. Measured in kilo pascal.

Ambient Pressure
The ambient pressure is defined as the pressure measured in the water at the diver's mouth level. Measured in kilo pascal.

Over/*Under Pressure*
The over/under pressure is the maximum pressure difference to which the diver may be exposed in the event of apparatus failure. It is measured relative to the ambient pressure and is the design maximum for emergency relief valves, shutoff valves and flow fuses. It is not a working pressure for normal operation. Measured in kilo pascal.

Respiratory Minute Volume
Respiratory minute volume is the mean volume of breathing mixture inhaled by the diver in one minute. Measured in litres per minute at ambient temperature and pressure.

Performance Goals

This section sets the performance goals which should be aimed at in Underwater Breathing Apparatus design.

Ventilation
The apparatus should be able to function satisfactorily over a range of ventilation defined according to its category

Work of Breathing
The work of breathing done by the diver, in using his breathing apparatus, increases with Respiratory Minute Volume. Work of breathing should be as low as possible. For test purposes, stan-

TABLE 4
Test conditions and performance goals for UBA evaluation.

Work rate	RMV (l/min)	Tidal volume (l)	Breathing frequency (breath/min)	V CO$_2$ (l/min)	Work of breathing (Joules/l)	DP (kPa)
Light	22.5	1.5	15	0.9	0.2	1.4
Moderate	40	2.0	20	1.6	0.6	1.4
Heavy	62.5	2.5	25	2.5	1.5	1.4
Extreme	75	3.0	25	3.0	2.2	1.4

dard values of RMV should be used. Performance goals are presented in Table 4 as a function of the RMV.

Maximum Respiratory Pressure
The maximum respiratory pressure should be as low as possible. For test purposes, standard values of RMV should be used. Maximum respiratory pressure should not exceed 1.4 kPa.

Maximum Lung Overpressure and Underpressure
If the equipment fails, the lung overpressure or underpressure should be as low as possible and should not exceed ±6 kPa.

Test Procedures

Controlled Parameters

Gas Supply and Return Pressures
According to manufacturer specifications.

Breathing Gas
Bottom breathing mix as defined in diving instruction for the depth considered.

Ambient Temperature
4 to 7°C water temperature.

Environmental Simulation
UBA set in vertical position, immersed, with at least 50 cm of water above diver's mouth level.

Test Depths
From minimum to maximum depth of operation of diver with reasonable increment (50 m for instance in a deep heliox test).
  In cases where the breathing performances of the UBA are dependent on the diver's position relative to the bell, each test depth should be repeated with the bell above, at level and below the dive according to the minimum and maximum allowed excursions.

Divers Simulation
Diver represented by a manikin with 'human like face'. Airways should be simulated by at least 4 cm internal diameter piping.

DP sampling tube should be located at the diver's mouth end, perpendicularly to the pipe.

No artefact should be used for sealing an oronasal mask on the manikin head face.

Breathing Machine

Sinusoidal waveform with an inhalation and exhalation ratio of 1 (5% allowable variation). Dead space due to breathing machine pipe connexions should be reduced to minimum. Its volume should be measured and indicated in the test report.

Ventilation

See Table 4 for standard values of tidal volume and breathing frequency.

Two Divers/Test

In a case where equipment is to be tested for two divers, measurements and tests should only be performed on one UBA.

The second diver should be simulated by a continuous gas flow controlled the same way as for the test ventilation on the first diver (i.e. same mean gas flow in litres per minute).

Measured Parameters

- Supply and return pressures (recorded versus time)
- Any relevant pressure and pressure drop (recorded versus time)
- Breathing resistance DP (recorded versus time and piston movement)

Computed Parameters

Work of breathing computed from DP/V loop by mean of digital integration or plannimetry techniques.

Test Equipment

Environmental Simulation

- One chamber with controlled temperature wet pot for diver simulation.
- One chamber for bell simulation.

Diving Equipment

- Surface breathing gas panel.
- Bell umbilical.
- Bell breathing gas panel and auxiliary systems.
- Diver umbilical.
- Helmet, gas heater, bail out system if required.

Diver Simulation

- Manikin head.
- Breathing machine.
- Water drain for helmet.
- Eventually, a electrical motor for manual adjustment of regulator.

Transducer

- P and DP transducers

Recorder

- Strip chart recorders for pressures and DP.
- x/y table for DP/V loop.

Second Diver

- A controlled gas flow meter to simulate the second diver.

INSPIRED GAS TEMPERATURE STANDARDS

Performance Goals

Ventilation
Inspired gas temperature should be measured using a range of standard ventilations defined according to equipment category.

Minimal Inspired Gas Temperature
Minimal safe and comfort inspired gas temperature is defined, or heliox breathing, as a function of depth (Table 5).

TABLE 5
Minimal safe inspired gas temperature

Depth (m) }	Minimum inspired temperature (°C)		Comfort inspired temperature (°C)
150	10		23
200	15		25
250	18		26
300	20		28
350	22		29
400	24		30
450	25		30

Controlled Parameters
Same as breathing performance study plus:

- Hot water flow to gas heater, according to manufacturer specifications.
- Hot water temperature at bell level: 40°C to 42°C.
- Diver's umbilical, gas heater and helmet immersed in cold water: 4 to 7°C.
- Breathing machine heater:
  The breathing machine should be equipped with a heating system in order to reproduce gas rewarming inside the lungs. The exhaled gas temperature (Tex) should be related to inhaled gas temperature (Tin) according to a linear relationship:[8]
  Tex = 0.28 Tin + 25.4°C.

Measured Parameters

- Relevant temperature along gas and hot water circuits.
- Hot water temperature at gas heater inlet and outlet.
- Gas temperature at gas heater inlet and outlet.
- Inspired gas temperature. (All parameters recorded versus time).

Computed Parameters

- Gas heater output power.
- Gas heater input power.
- Gas heater efficiency.

Test Equipment
Same as in breathing performance study plus:

- Surface hot water machine.
- Bell hot water equipment (flowmeter and thermometer).
- Diver hot water umbilical.
- Gas heater.
- Temperature control system for wet pot.
- Temperature transducers.
- Strip chart recorder for temperature readings.

INSPIRED CO_2 LEVEL STANDARD

Performance Goals

Minimum Inspired CO_2 Level
The design of the UBA should minimize CO_2 level in the apparatus. The maximum inspired partial pressure of CO_2 should not exceed:

- 2 kPa peak value.
- 1 kPa average value.

Ventilation
The apparatus should be able to function correctly over a range of standard ventilations defined according to equipment category.

CO_2 Injection Rate
In order to reproduce physiological CO_2 production, CO_2 should be injected into breathing machine at a constant rate equal to 4% equivalent of surface RMV. See Table 4.

Test Procedure

Controlled Parameters
As in breathing performance study.

CO_2 Measurements
Inspired CO_2 should be measured at diver's mouth level. Peak values should be measured using a mass spectrometer which is the only analyser that can provide a fast enough response.

Average values can be obtained from the mass spectrometer signal or from a classic infrared analyser because its sampling line introduces damping effects.

CO_2 *level* should also be measured:

- in helmet,
- in any relevant location (cannister inlet/outlet).

TEST PROTOCOL

To be used for:

- Breathing performance studies.
- Inspired gas temperature studies.
- Inspired CO_2 level study.

A — Chamber on surface.
B — Make sure that UBA is connected according to manufacturer specifications and functions correctly.
C — Calibrate transducers, analysers and breathing machine.
D — Check pressure reference system (if required).
E — Connect heliox bank and pressurize the system.
F — Run the system and analyse breathing gas.
G — Pressurize chamber system to first test depth.
H — Set breathing machine to 75 l/min ventilation and CO_2 injection system to 3 l/min.
I — Adjust reference pressure system (eventually).
J — Adjust controlled pressure.
K — Adjust regulator and purge the helmet water (eventually).
L — Adjust hot water flow and temperature at bell level.
M — Wait for steady state and take readings.
N — Set breathing machine to 62.5 l/min ventilation and CO_2 injection system to 2.5 l/min and repeat step M.
O — Set breathing machine to 40 l/min ventilation and CO_2 injection system to 1.6 l/min and repeat step M.
P — Set breathing machine to 22.5 l/min ventilation and CO_2 injection system to 0.9 l/min and repeat step M.
Q — Stop breathing machine, hot water flow and CO_2 injection.
R — Pressurize chamber system to the next test depth.

S — Repeat steps H through Q.
T — Repeat step R to S till the maximum test depth.
U — Stop the system, breathing machine, CO_2 injection and hot water flow.
V — Decompress chamber system to surface.
W — Chamber on surface, dismantle and check UBA.
X — Check transducer calibration.
Y — Check pressure reference system.

Appendix II

RELIABILITY STANDARDS

Definition

Failure
A failure is defined as any change in the system that modifies the breathing performances of the UBA.

Maintenance
No maintenance should be performed and no parts should be changed during the reliability test.

Performance Goals

Mean Time Between Failures
A mean time between failure of 160 hours with a confidence level of 95% is to be demonstrated. It represents a reliability of 0.975 for a four hour mission which is taken as the average bell lockout time.

It represents about 480 hours of test without any failure. However, if the system is proved to have an emergency back-up for each function (bail out system, bell shuttle valve, etc.), the test time can be divided by 2, i.e. 240 hours without failure.

Ventilation
Reliability study is to be performed using 40 l/min ventilation.

Test depth: Maximum depth of operation.

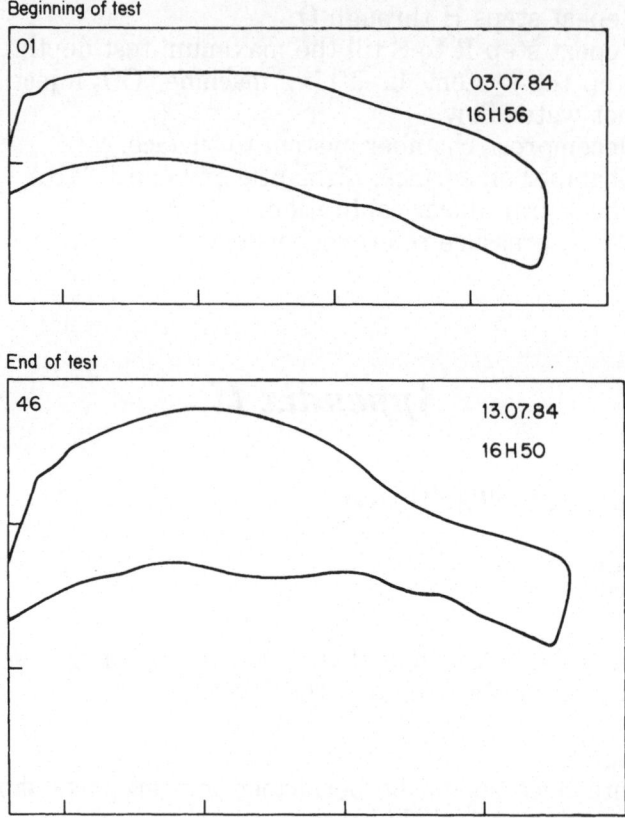

Fig. 4.4 Reliability study of the Comex gas reclaim system according to the procedures described in Appendix II. Results presented were obtained after 240 hours of continuous test at 450 m using a 45.5 l/min ventilation. Diagram displays the DP/V loop at beginning and end of test

Test Procedure

Controlled Parameters
Same as in breathing performance study, except that test temperature is ambient temperature.

Measured Parameters

- Breathing resistance at diver's mouth (DP/V loop).
- Breathing machine piston movement.
- Supply and return pressures (recorded versus time).

- Any relevant pressure and pressure drop (recorded versus time).

Test Equipment
Same as in breathing performance study.

Test Protocol

A — Chamber on surface. Check that equipment is connected according to manufacturer specifications and functions correctly.
B — Calibrate transducers and breathing machine.
C — Connect heliox gas bank and pressurize system.
D — Run system and analyse breathing gas.
E — Pressurize chamber to test depth.
F — Set breathing machine to 40 l/min ventilation.
G — Adjust reference pressure system (eventually).
H — Adjust controlled pressures.
I — Adjust regulator and purge helmet if required.
J — Wait for steady state and start reliability trials.
K — After 240 hours of continuous running, stop breathing machine and system.
L — Decompress chamber to surface.
M — Dismantle equipment and check parts for damage or wear.
N — Check transducers calibration.

5

Hyperbaric Welding Habitats: Environmental and Safety Aspects

C. F. Lafferty RN, MNI, British Gas Corporation

INTRODUCTION

The aim of any hyperbaric welding operation must principally be to provide the client with a weld which satisfies the client's requirements and standards. A habitat is only provided to enable a weld of the right quality to be made and to provide a safe working environment for the welders.

The technology of producing a weld to the required standard is well documented and is not discussed here. I will examine some of the conflicts which arise from the need to satisfy both the production of a satisfactory weld and the provision of a safe environment.

To do this I intend to examine two typical welding operations. One is a structural repair/strengthening operation and the other a pipeline tie-in operation. Both are based on actual North Sea operations and both are in relatively shallow waters.

Shallow waters provide some specific problems which are not apparent in deeper operations. Both examples have been altered to protect the actual operators from the public gaze and neither is used to imply any critiscism but to enable discussion based on real rather than theoretical conditions.

63

OPERATION A

This task is a node strengthening exercise undertaken in about 30 feet of water. The operation consisted of welding large steel chord wraps to each of the nodal members. It was a very complex welding operation requiring very high levels of pre-heating and very large weld deposits to be made.

Because of the structural limitations and time constraints the habitat was small and very congested. With the chord wrap material, the myriad of pre-heating mat tails, the divers' equipment, the welding equipment and other necessary tools the inside of the habitat resembled a tropical jungle. The atmosphere was compressed air supplied from the surface support vessel. The temperature and humidity were very high, approx. 150°F, when pre-heat was being applied.

Compressed air was used because of its simplicity and because of the welding pre-qualification procedures previously agreed and tested.

The diving procedures used were that the divers were deployed in a cage wearing dry suits and helmets. They swam to the habitat and entered through the bottom habitat entrance. Inside they would remove their diving equipment and don flameproof overalls. When pre-heating was being applied the divers also wore tube-suits and plugged these into a cold water supply piped to inside the habitat. This enabled the welder/divers to operate in shifts of up to 8 hours. Without it they were becoming exhausted after a very short time. (See Fig. 1.)

Habitat atmosphere was controlled by flushing through as necessary and gas samples were analysed at the surface. The welding used was the conventional Manual Metal Arch technique. A considerable amount of grinding was required but the

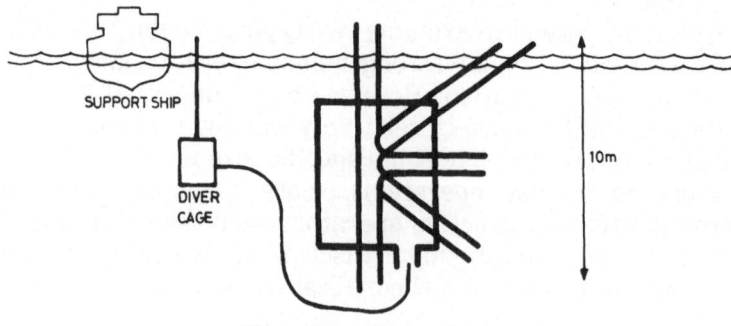

Fig. 5.1 Operation A

divers wore lightweight breathing masks during the welding operations when noxious gases were being generated.

The selection of the welding procedures and the choice of a compressed air habitat atmosphere were based largely on economics and satisfactory precedents in offshore operations.

Problems Encountered

The first problem was the very limited space available to the diver/welders resulting in difficulty in stowing equpment, difficulty in moving quickly to the entrance and difficulty getting in and out of the entrance, which was also used as a cable entry.

The divers had to remove their bulky dry suits and diving hats and stow them out of harm's way. They then plugged themselves into the cold water supply and when welding was being done they donned their breathing masks. This gave them a rather unwieldy umbilical to manoeuvre around the habitat. It also meant that there were large quantities of inflammable materials present in the habitat, which were a potential fire source. The compressed air atmosphere provided a very high fire risk having both a high oxygen supply as well as a high oxygen partial pressure. In addition the use of welding and grinding equipment provided a genuine fire source.

The problems of fire in hyperbaric situations is quite well documented but is surprisingly not widely understood.

I will briefly summarize the problems. Fire even in hyperbaric situations requires O_2 as one of its three necessary elements, fuel, heat and oxygen, to initiate or sustain it. The higher the oxygen percentage in the gas supplied the easier it is to initiate a fire. It has been shown, however, that to initiate fire in hyperbaric conditions the O_2 percentages in the gas supplied must be greater than 8%. This is regardless of the partial pressure of the O_2 present in the atmosphere. As the percentage of gas increases above 8% the risk of fire initiation increases (Fig. 2).

On the other hand the partial pressure of gas affects the rate of burning as well as the propagation, once initiation has occurred. A high partial pressure of O_2 will increase the rate of burning.

Thus in fire prevention terms it is necessary to try to reduce the percentage of O_2 supplied as well as to reduce the partial pressure present in the habitat. To prevent the possibility of fire ever occurring the O_2 supply must be kept below 8%. This may not always be possible in shallow water, due to the need to sustain human life. (I discuss this in more detail later.) In deep water, O_2 contents below 8% are often the norm and therefore fire is more of a problem in shallow water operations.

These problems were recognized but expected to be counteracted by good housekeeping. The precedent, as I have said earlier, was favourable. Despite the documented theory, common practice seemed to indicate that the problem was not serious.

In fact several fires of various intensity did occur on this operation. The more serious ones caused quite severe damage to equipment and required the occupants to evacuate the habitat at speed.

The fires highlighted several related problems. In the congested area the occupants had difficulty reaching the exit and in the smoke this was an increased hazard. The welders had to remove their lightweight masks and find their diving helmets and diving suits. In fact there was no time to don any equipment and on at least one occasion the diving helmets/equipment were on fire, and not usable. The welders therefore had to jump into the water with no breathing equipment nor thermal protection. On one occasion one of the divers got halfway to the surface and realised he was till connected to the habitat cold water supply. Still on a deep breath, he had to return to the smoke filled habitat, enter it and disconnect himself. The escaping men, without breathing equipment, were obviously risking embolisms in making emergency free ascents and on reaching the surface they were not attached to any safety line or umbilicals.

Even after making improvements in housekeeping and other procedures fires occurred. On this occasion precedent seemed to have little to offer.

Possible Solutions

A variety of solutions were considered both as short term and long-term answers:

1. Change the Atmosphere to a Nitrox Mixture

This would have the advantage of reducing the O_2 supply and thus reducing the risk of fire initiation. Because of the shallow depth and the (debatable) need to maintain a life supporting atmosphere, it was not possible to reduce the atmosphere to

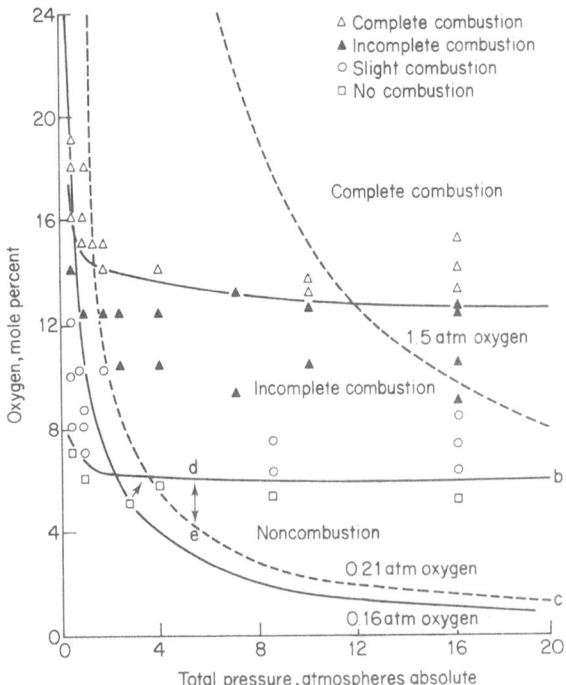

Fig. 5.2 Three combustion zones for vertical paper strips in N_2/O_2 mixtures, showing in particular the zone of no combustion (from Cook *et al.*, 1968, as reproduced in the *Underwater Handbook*, Shilling *et al.*, 1976)

below 8%, and so there was a risk of fire occurring; though of course a reduced risk.

The method of controlling a nitrox atmosphere and the need to supply unusual gas mixtures to the site raised questions of costs, complexity and practicality. In this particular case it also may have necessitated re-design of the habitat.

More immediately critical was the welding procedure. The procedure had been qualified in the compressed air environment. A change in the nitrogen content brought in welding related problems which would affect the primary aim, i.e. the production of a high quality weld.

2. Change to He/O$_2$

This is always a possibility and of course in deep water it is the normal practice. In this water depth the use of He/O$_2$ would bring in diving related problems such as reduced shift times, decompression problems and increased embolism problems, which require careful evaluation and possibly extensive investigation or even developments.

The use of He/O_2 would necessitate the provision of sophisticated and expensive atmosphere control equipment as well as the provision and expenditure of expensive and special gas.

The welding procedures of course would have to be requalified and in this particular depth He/O_2 was not considered a viable solution.

3. Change to Argon/O_2

The incentive to consider this course is that the use of argon would be less likely to affect weld quality.

As with the other two solutions discussed above, atmosphere control would be more complex. Additionally the use of argon in anything other than experimental diving work in the past discouraged serious consideration of this solution.

In the three solutions discussed above all have a common link. Is it necessary to provide a breathable habitat atmosphere?

Is it acceptable to provide an almost completely inert atmosphere or at least one with an O_2 supply percentage below the magic 8% mark?

An inert atmosphere would ensure that no risk of fire initiation ever occurred. It would, however, entail the welder having to work in a poisonous atmosphere and rely on his breathing apparatus to safeguard him. Some people have taken the view that this is no different from the diver in the water and is therefore acceptable. My own view, however, is that this is introducing a known hazard and therefore simply substituting one risk for another. I would, however, seriously consider reducing the O_2 content to a minimum breathable level, e.g. 0.16 atm oxygen, to sustain life if the mask had to be removed for a very short period. This would help to reduce fire risk without introducing any other risks.

Of course, if a low O_2 atmosphere is present in the habitat and a higher percentage of O_2 is being breathed by the diver then steps to prevent O_2 build up in the habitat must be taken.

4. Provision of a Safe Haven

This is not a solution to the atmosphere problem but a possible solution to some of the other problems.

A Safe Haven here could be a small additional compartment or even a bell attached or close enough to the habitat exit to avoid a wet transfer. Such a compartment could be used to stow the diving equipment and thus reduce the inflammable material in the habitat. If the Safe Haven is provided with lights it can also be a refuge into which escaping divers can be accommodated. This would have the advantage that divers could dress into their diving equipment, they could communicate with the surface and they could ascend to their support vessel using their normal routine procedures. It is an obvious disadvantage that this solution increases the size of the habitat.

Notwithstanding I would suggest this is a facility that should always be provided when wet transfers from the surface are necessary.

5. Improve Welder/Diver Clothing

The clothing provided to the welders is sometimes inadequate. The overalls should be flameproof, as should the footwear, headgear, etc. If the man has to enter the water in an emergency then his clothing should be capable of protecting him from thermal shock for the expected period of his immersion.

6. Improve Welder/Diver Breathing Equipment

If the welder must wear a breathing set inside the habitat then that equipment should be at least flame retardant or resistant. The equipment should be suitable for transfer either to the safe haven or to the surface if no haven is provided. There should be no need to remove or replace breathing equipment before escaping from the habitat.

7. Emergency Drills

Many of the problems experienced during a real emergency, such as the enforced evacuation of a fire filled habitat, can be eliminated by the occasional simulation of an emergency. For instance the need to disconnect the cooling suit I mentioned earlier would have been obvious if a drill had been done.

Emergency drills are unpopular because they cost time and money. One argument against them is that if a job is well planned, well organized, well managed and well manned then emergencies will also have been considered and the action to be taken will be automatic. Therefore drills are unnecessary!

Another argument is that the conduct of a simulated emergency can put personnel into an unnecessary risk situation.

However the use of training and drills is well proven as a reliable way of reducing risk to both projects and personnel. In cases such as this, which really can be considered as 'one-off' operations, it should not be too difficult to include some drills into the project planning and to obtain the client's agreement.

Actual Solution

I have discussed various options that were considered but the actual solution selected for the next year's work was the use of a cofferdam.

The cofferdam has the advantage of making use of only 1 atm of air and thus from both welding quality, fire prevention and atmosphere control the problems are as near as possible those familiar to anyone welding on the surface.

Some of the other problems faced in cofferdam operations of course are the ability of the structure to sustain the cofferdam in the expected environmental/weather conditions; the access to and from the working compartment and the installation and removal expenses involved.

The cofferdam solution is very attractive but will have its limitations in both depth and acceptable structural loadings.

OPERATION B

This operation is similar to Operation A in that a high quality hyperbaric weld was required to be done but very different in most other aspects.

The hyperbaric welding was being done to complete a 36-inch pipeline installation in about 120 ft of water.

The habitat was installed over a pipeline and the entrance way was dug out. To provide adequate welding shift times saturation diving techniques were being used and the habitat atmosphere was 20% O_2 and 80% He.

The divers transferred from a conventional bell and did a wet transfer to the habitat. During the transfer the divers wore dry suits and diving helmets which were removed and stored inside the habitat.

The divers entered the habitat through a small transfer compartment and a hatch, which they closed behind them. During

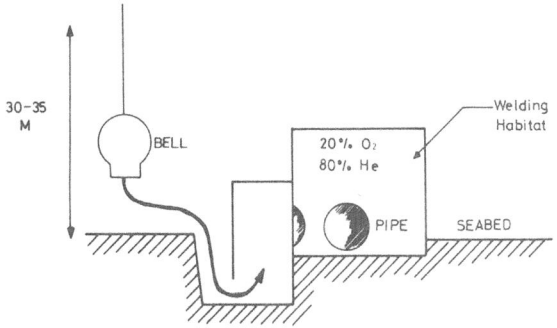

Fig. 5.3   Operation B

welding operations the divers were not required to wear separate breathing equipment as the habitat was provided with a full atmospheric control system operated and monitored from the surface.

The welders were normally more familiar with operating in a dry-transfer mode. That is, the bell is normally locked on to the habitat to allow the divers to transfer to the habitat without entering the water. The bell in this way also provides them with a perfect safe haven. Normally therefore the welders have been selected for their welding skills and although qualified as divers they do not need to be very experienced in the water.

Possible Problems

The habitat atmosphere selected had a high O_2 percentage and an extremely high O_2 partial pressure. Thus an excellent fire initiation atmosphere was provided as well as an excellent fire propagation atmosphere. The welding procedure had been approved in this atmosphere and any changes considered after the qualification procedures would impinge on these procedures.

As in the previous example the divers had to leave their infalmmable diving equipment inside the habitat, causing both congestion and a fire hazard. The space available inside the habitat was very adequate.

In an emergency evacuation the divers would have had to either take the time to put on their diving equipment or more probably jump into the cold water without any equipment. In saturated conditions the divers must not ascend to the surface so they must swim to their bell. In this case the bell was only about 15–20 ft from the habitat exit but bottom visibility conditions were poor. As mentioned earlier the divers were not very experienced and therefore were at higher risk in these circumstances.

In fact there were no untoward incidents in the course of this operation and therefore there can be little justifiable evidence of undue risk to personnel.

Possible Solutions

As there were no actual emergencies then the discussion of solutions may seem academic. I think though that each operation can be assessed both before and after completion and lessons learned.

In this case I would like you to consider the following:

Divers' Experience
The welder/divers were being asked to operate outside their normal dry transfer procedures. There is surely a case where consideration should be given to maximizing the diving experience or confidence of the welder/diver.

Reduction of O_2 Content
As previously discussed the O_2 content could in this case have been reduced to close to or even below the level required to prevent the possibility of fire ever occurring. The penalty for that would be an increase in the He content and this would need to be investigated as to its affect on weld quality. In fact in this case it was found that if pre-heating levels were raised 5°C the weld quality remained acceptable.

Safe Refuge
If the entrance lock on this chamber had been enlarged it could have been used as a dressing station and safe refuge as we previously discussed. This would have avoided the possibility of unprotected divers having to enter the water. In addition, in this case, it would have enabled the divers to remain in communication with the surface and avoid any emergency or panic measures being taken.

Emergency Drills
If emergency drills had been conducted at the start of the operation any undue delays in transfer from the habitat to the bell or any local difficulties could have been identified.

Actual Solution

As no incidents occurred and as no immediate repeat of this particular operation was planned by the client no long term solution was needed. But because of various commercial pressures no short term solutions were implemented either and although the opportunity existed it was not taken. As no incidents occurred perhaps the fact that no action was taken on these points was eventually justified. I hope though that in future operations these considerations will be given greater thought.

CONCLUSIONS

The decision on what habitat atmospheres and conditions to use must take serious consideration of:

(a) Weld procedures
(b) Life support requirements
(c) Fire
(d) Emergency evacuation
(e) Habitat location

I conclude that in shallow water operations some of these items are not always thought through completely and would recommend that the following should always, at least, be considered:

(a) The provision of a low O_2 gas supply with a percentage below 8% where possible
(b) The maintenance of as low a ppO_2 as is possible compatible with a safe breathing atmosphere. I would suggest a ppO_2 of 0.16 atm where short exposure is expected
(c) The provision of good atmosphere control and monitoring
(d) The provision of a safe refuge
(e) The exercise of emergency procedures
(f) The improvement in welder/diver clothing and breathing equipment.

REFERENCES

1. *Underwater Handbook*, Shilling *et al.*, 1976.

2. *Hyperbaric welding on compressed air atmospheres*, Reims Consultants.
3. *Hyperbaric Fires*, Comex Services.

6

In-water Secondary Life-support Systems: Removing the 'Emergency' from Incidents

N. V. Sills, Consortium Salvage Ltd.

SUMMARY

This chapter describes what is meant by the term 'life-support' in the context of hyperbaric diving and examines the reasons for and requirements of secondary life-support systems, sometimes called bail-out systems. Equipment and some of the operational aspects, including procedures, training and ancillary items are discussed.

An opinion on the level of secondary life-support required in shallow, medium and extreme depths is presented to provoke discussion and help identify potential solutions for some of the remaining problem areas. Primary life-support systems are only discussed where background information is relevant.

INTRODUCTION

In hyperbaric diving operations, life-support is the term given to equipment that supports the function of biological processes in an otherwise lethal environment, that is:

- respiration;
- maintenance of body temperature;
- control of environmental pressure.

In saturation diving, facilities must also be provided for:

- nutrition;
- excretion;
- sleep.

Failure of the system to provide any of these facilities will, after an initial 'buffer' period, when the human body depletes its built-in storage capacity, or its physiological tolerance to change is exceeded, lead to severe distress and ultimately death.

In commercial diving operations, exposure to a non-life-supporting environment, the in-water work period, is normally less than 8 hours. It is, therefore, only those requirements with a 'buffer' or tolerance period shorter than 8 hours that are normally provided for by the life-support system, that is breathing gas, maintenance of body temperature and control of environmental pressure.

In the case of in-water incidents, a period of 10–15 minutes is usually considered adequate to allow a diver to return to a more permanent 'dry' life-supporting environment (the surface, a submarine or a diving bell) if the primary in-water system fails. It is, therefore, only those requirements with a 'buffer' period shorter than 10–15 minutes that are provided for by secondary life-support systems. In shallow water, this is usually breathing gas and a method to control environmental pressure. In deeper water, as pressure increases and ambient temperature decreases, the 'buffer' periods are reduced due to a complex interaction of physical and physiological factors. A point is reached where the secondary system must also maintain body temperature. As depth increases, it also becomes more difficult to design and build secondary life-support systems with an endurance of 10–15 minutes.

An indication of the length of the 'buffer' periods before unacceptable distress sets in can be gained from the following examples. The times are necessarily approximate.

Situation 1

Diver on holiday in warm climate, on land and at rest

Requirement	*Buffer period*
Breathing gas	<2–3 minutes (limit of breathholding)
Control of temperature	not required

Control of pressure	not required
Food	>7 days
Fluid	>3 days

Situation 2

Shallow air diving in warm sea and undertaking moderate work

Requirement	*Buffer period*
Breathing gas	<1–2 minutes
Control of temperature	<6–8 hours (without body insulation)
Control of pressure	<20 metres per minute decrease
Food	<3 days
Fluid	<2 days

Situation 3

Deep diving using mixed gas and in cool sea, undertaking moderate work

Requirement	*Buffer period*
Breathing gas	<1 minute
Control of temperature	<2 hours (with body insulation)
Control of pressure	<20 metres per minute within excursion limit
Food	<1 day
Fluid	<1 day

Situation 4

Extreme deep diving using mixed gas and in cold sea, undertaking moderate work

Requirement	*Buffer period*
Breathing gas	<30 seconds
Control of temperature	<5–10 minutes (with body insulation)
Control of pressure	<20 metres per minute within excursion limit
Food	<1 day
Fluid	<12 hours

DISCUSSION

For the purposes of this discussion, I am going to define an incident as an event, or series of events, that remove the primary life-support facilities from a diver in the water, forcing him to become autonomous.

The title 'In-water Secondary Life-support Systems: removing the emergency from incidents' serves to imply that secondary systems should be regarded as complete escape systems, rather than a simple 'bail-out bottle' that supplies breathing gas. I would like to labour this point by describing an incident.

Imagine a diver working from a diving boat tied up alongside a ¼ million ton oil tanker which in turn is moored from a single point on the bow. The diver is using surface orientated air diving equipment, including a secondary life-support system. The diver is undertaking a television survey of the tanker's hull. The water is warm and clear, with little current at the working depth of 16 metres, although the bottom, some 12 metres below, is hazy and the features indistinct. The diver becomes aware that he is being dragged slowly towards the seabed, but from the starboard side of the tanker, not the port side, on which the support boat is tied up. He asks for more slack on his umbilical, only to be told by the Diving Supervisor that it is nearly all out and that he is having to make if fast. At this point, the diver can no longer hold on to the marker rope lashed under the tanker's hull as he is being pulled downwards and to starboard. Suddenly aware of his predicament, the diver activates his secondary system (an air bail-out), grabs his knife and hastily starts cutting through his umbilical. The diver felt that his 'buffer' system would be hard pressed to bring him from under the tanker to the surface. At this point, the diver sees he is being dragged towards a large mass of scrap on the seabed. He cuts the last strand of his umbilical and watches the cut end disappear into the tangled pile of scrap. The diver swims back under the hull to the support boat.

An analysis of the incident showed that the tide had changed and the tanker had swung on its mooring with the support boat tied to it. The tender had allowed too much slack on the diver's umbilical and it had fouled on the pile of scrap on the seabed. The diver, concentrating on the survey, had not checked his umbilical.

If the diver had been deeper, say 50 metres, and outside the 'no-stop' curve, the loss of surface communications, and thus depth information (pneumofathometer), would have robbed him of a means to effectively control his environmental pressure (ascent). For this, he would need surface independent equipment to gauge water depth/time, and compute ascent rate. If the incident had occurred at night, a location device, such as a flashing strobe or torch, or in misty conditions a whistle, would have been used. In choppy seas, a life jacket would be important to keep his head above water.

It is true to say, however, that in many circumstances the standby diver would attend the scene and provide some of the necessary ancillaries. Also, only a partial failure of the primary system, say breathing gas, may occur, leaving the other services and facilities intact.

What then constitutes a diver's in-water secondary life-support *system*?

If one examines the given incident, it can be seen that there is a distinct sequence of events:

- development of the emergency;
- the diver's appreciation of the situation;
- the decision to go to secondary life-support and activation of it;
- the use of tools to release the primary system/effect escape;
- orientation towards the surface;
- controlled return to surface.

From this brief analysis, it is possible to construct a basic model of a secondary life-support system (or escape system) which can be applied or adapted to almost any depth or circumstance. The analysis also draws attention to one very important part of the model — the diver himself. He has to assimilate information, make decisions and act on them if his escape is to be successful. This, of course, is the reason why proper training and established procedures are so important. Although not a tangible piece of equipment, they form an integral part of the secondary life-support system.

The basic hardware for the model is a self-contained, diver carried and activated system with an endurance of 10–15 minutes that supplies:

- breathing gas at a suitable pressure and flow;
- information to allow control of environmental pressure (depth/time/tables);

- tools to aid escape (knife);
- orientation to allow return to a life-supporting environment (this may be a visual cue, a shotline, direction-finding instrument, etc.);
- location device to allow surface team (or bellman) to locate the diver (strobe, whistle, etc.);
- buoyancy aid (on surface only).

Naturally, the equipment should be easy to put on and take off, be comfortable to wear and neutrally buoyant in water. Its buoyancy should not change appreciably during use.

How the secondary equipment can be integrated with the primary equipment, and what parts of the primary system are treated as infallible, are touched on later.

If one applies this basic model to mixed gas bell bounce and saturation dives, say to 120 metres, a small number of differences become apparent.

Since the diver does not and, indeed, cannot return directly to the surface, but goes instead to a diving bell or submarine, he does not have to control his in-water ascent (or descent) so precisely — the relative pressure changes are smaller. Also he would normally be less than 30 metres from the bell (length of his umbilical). However, the diver's target, the bell trunking, is very much smaller than the surface of the sea and its small diameter restricts the size of equipment he can wear. Where a diver has to release his umbilical in poor visibility, he may find it difficult to re-locate the bell. A homing divice, such as a messenger line or an acoustic torch, should be considered part of the system.

It is likely that the diver would be wearing equipment to maintain his body temperature in these forms of diving. However, a failure of the primary system, often hot water suit and gas heating, is considered of little consequence in the time frame of an incident. It is extremely unlikely that the suit's thermal insulation properties would be lost — the diver would have to remove his suit. Even if this were the case, the diver's 'buffer' system, shivering and vasoconstriction, would give him a sufficient measure of protection. A large compressed gas cylinder and demand valve integrated with, or part of, his primary breathing circuit is sufficient for 10–15 minutes of autonomy.

Efforts are now being made to extend the operational depth of diving to 400 metres. Here the 'buffer' systems of the body are reduced so much that a new biological requirement must be introduced to the secondary life-support equipment if it is to maintain the diver in an active state for even 10–15 minutes. That is gas heating.

The loss of heat caused by breathing cold gas at this depth is so great that respiration can become inhibited within minutes and incapacitate a diver to such an extent that he may be unable to effect his own return to the bell.

The technology that is used to provide a secondary supply of breathing gas in shallower depths, open circuit compressed gas, cannot be applied. The endurance of the largest practical compressed gas unit would be only 1 or 2 minutes. Furthermore, demand valves designed and built for shallower depths have gas delivery characteristics unsuitable for emergency use at 400 metres.

Due to the effects of pressure and density, a diver's reactions are often seen in experiments to be more sluggish and less precise at this depth. The diver may, therefore, find it difficult to operate the controls of a 'conventional' secondary life-support system.

It is clear that the equipment designed for shallow and medium depths, though similar in principle, would, in fact, be of little use at 400 metres. It is equally clear that the loss of a primary in-water life-support system during a dive at this depth will lead to an emergency in a very short time. To prevent this, any secondary system must:

- restore breathing gas within 30 seconds;
- restore gas heating within a few minutes;
- prevent excessive body heat loss.

It should have an endurance of 10–15 minutes and be able to orientate the diver to allow his return to the bell in bad conditions. Due to the increased complexity of the system, and the more rapid response required in an emergency, consideration should be given to activating the system by one simple operation. (Appropriate measures would need to be taken against accidental operation.) Provision may also have to be made for release of the diver's umbilical in the event of fouling.

At the present time, there is no commercially available system capable of providing these facilities.

There would appear to be at least three options around which to design a system.

(1) A closed or semi-closed circuit rebreather. Diver-carried.
(2) An open or closed circuit emergency unit deployed at the work site prior to the dive and which the diver 'plugs' into on arrival.
(3) An open or closed circuit system mounted on a remotely operated vehicle or manned submersible and deployed at the work site.

Although option (1) is the most attractive of the three, at the present time, it is perhaps the most difficult to design and build due to the requirements of small size, simplicity and ruggedness. In practice, it may also prove difficult to integrate a closed circuit demand apparatus with a free flowing closed circuit gas recovery unit because of large 'dead' spaces, pressure differentials and complex changeover controls.

Part of the solution may lie in an idea suggested by Professor Lambertsen a few years ago which is based on the observation that 90% of the carbon dioxide expired by man is expelled in the last 10% of each exhalation. If a device could be developed to exhaust the last 10% of each exhalation, but retain the remaining 90% for rebreathing, then the size, improved breathing characteristics and reduced complexity of such a semi-closed circuit rebreather might favour a diver-carried apparatus. Much of the dead space and resistance to gas flow would be lost with the elimination of carbon dioxide absorbents. Experiments show that inspired gas temperature at depths equivalent to 500 metres should be at least 25°C. this can be achieved by using passive heat exchangers to transfer heat from expired gas to inspired gas, provided the 'dwell' time of each slug of gas is sufficient. This principle does not work effectively on a free flowing system unless the circuit is small and very well insulated. Active heating could be provided for a short time by using battery powered autotherming trace heaters of the type manufactured by Raychem. Raychem is, in fact, developing and electrically heated suit using this material at the present time.

Option (2) would be comparatively simple to design and build, and because size would not be a particular problem, either open or closed circuit principles could be adopted. However, the practicality of deploying such a system from the surface and the limitation placed on the diver in terms of work footprint might be too restrictive. A procedure whereby the diver leaves the bell with no secondary life-support until he reaches the work site and has to plug in an ancillary umbilical may well be entirely unacceptable. Likewise, a return to the bell 'uninsured' is not satisfactory. A compromise might be for a secondary system too large to be worn, but still portable, to be carried on the outside of the bell. The diver would carry it from the bell and position it with him on site. He would again be connected to it by a short umbilical. However, this might prove awkward and time consuming in mid-water tasks.

Option (3) would be comparatively simple to design and build, but would only be practical on the larger ROVs and manned submersibles. The ROVs used to monitor and assist divers at present are the smaller, less expensive machines; large ROVs could represent a hazard to diving operations and are costly.

In these extreme depths, however, the advantages of having a vehicle capable of supporting a diver in an emergency and being able to escort or carry him back to the bell might be persuasive, especially if the ROV or submersible also carried the diver's power tools and other aids for tasks. Additional advantages would be the possibility of alternative communications to the surface and the bell via the ROV or submerisble and the psychological benefits of having a 'buddy' present. In the case of a severe injury to a diver, the ability to transport the casualty back to the bell may be important.

In both options (2) and (3), there would be a necessity to 'plug in' a secondary umbilical and this would require comprehensive changes, not only in procedure and training, but also in the law.

CONCLUSION

I would like to conclude by suggesting that in the event of in-water primary life-support failures, most potential emergencies could be downgraded to incidents rather than up to accidents if *full* secondary life-support systems are always provided. Most important of all is to consider the system as a combination of procedures, training, equipment and tools, rather than just a set of bail-out cylinders.

In extreme depths, the same basic philosophy applies. However, which of the three options cited will actually be adopted and for what reasons must surely be a subject for discussion. Perhaps there is even a fourth option for tasks on future very deep installations — built-in emergency facilities for the diver.

7

Breathing Resistance: Keeping the Requirements Realistic

K. Segadal, D. M. Furevik and E. Myrseth,
NUTEC (Norwegian Underwater Technology
Center), Bergen

SUMMARY

Breathing resistance is usually defined by work of breathing per volume respiratory ventilation. It is divided into internal and external resistance and also into an elastic and a resistive component. Total breathing resistance is an important factor, limiting a diver's physical capacity. In unmanned tests of breathing equipment, only external breathing resistance is measured. The ultimate and ideal requirement for breathing resistance is that the equipment adds no extra resistance to breathing, but instead assists respiratory ventilation so that the density dependent increase of internal resistance is reduced. Realistic requirements, however, must be set so that total cost to fulfil them is justified by gain in safety and efficiency. To determine this, breathing resistance must be weighted against other important properties of the life support equipment such as complexity, fail frequency, maintenance, mobility, entanglement, noise, vision, temperature, pCO_2, pO_2 and humidity. Requirements for breathing resistance are included in different proposed performance goals for breathing equipment. At a workshop held at NUTEC in 1981, Morrison's performance goals proposed for Department of Energy gained most support. NEDU also presented performance goals which were about equal on the important points. Main differences were: NEDU proposed only recommended, not also acceptance levels like Morrison, NEDU was more stringent for low respiratory ventilations and less stringent for high ventilations, hydrostatic imbalance (HI) was treated differently. The main problem is to interpret 'recommended' and 'acceptable' levels for breathing resistance. There should be no justification to use equipment not within the acceptance limits. Justifications to use equipment within acceptance levels, but not within recommended levels are: it is impossible to reach recommended levels and at the same time fulfil other requirements, important for safety and efficiency, the depth is shallow and working conditions are specially safe and easy. Today, with improved equipment and much deeper depths, 'realistic' requirements have come much closer to 'recommended' levels than they were in 1981.

NOMENCLATURE

WOB Work of breathing
UBA Underwater breathing apparatus
HI   Hydrostatic imbalance; the pressure difference between relaxation or reference pressure at mouth opening for UBA/lung and a specified pressure fixed to a location of the body, e.g. lung centroid.
msw Pressure as depth in metres of sea water

INTRODUCTION

The old acceptance criterion for breathing equipment was to put it on a diver and see if it 'worked'. If the diver was experienced and used to an inferior equipment, the acceptance was granted. For a long time this method was good enough, mostly because other aspects of diving techniques were more in need of improvement than breathing resistance. For several years now, however, the breathing equipment and its resistance have been recognized as one of the most critical factors in deep diving. A natural consequence of this was that a more precise and objective method for evaluating breathing equipment was needed.

For many years now, breathing resistance has been measured both on breathing simulators and on man. This has produced numbers that tell something about each equipment and makes it possible to compare different equipments. It is logical that there at the same time has been a wish to compare those numbers against accepted standards. The problem then was how to produce a standard. For a long time there were no real facts to base such standards on. It was at the same time felt that if a standard was set as it should be, no existing equipment would fulfil it. Nevertheless standards were proposed and used and at the same time more information was collected. This focused the attention on the characteristics of the breathing equipment and probably accelerated the developments in this field.

At the end of the 1970s there was general agreement that the breathing resistance would be one of the most crucial factors limiting operational diving depth. The need to agree on a common standard for breathing resistance thus became accentuated. On this background the Norwegian Petroleum Directorate and the British Department of Energy together arranged a workshop to achieve this. The workshop was held in Nov./Dec. 1981 at NUTEC in Bergen and gathered together the majority of the expertise in the field. A certain consensus for a proposal put forward by J. B. Morrison for Dept. of Energy[1] was achieved and summed up by the chairmen of the workshop.[2] Since then NPD and Dept. of Energy have carried on the work on this basis to produce a common guideline for breathing equipment performance.[3] In the workshop Middleton and Thalman presented the NEDU performance goals,[4] which were a little bit different on some points.

Later we will discuss if those proposed performance goals should be used as requirements, but first we will give a little explanation of what is meant by breathing resistance and why it is important.

BREATHING RESISTANCE

In fluid dynamics, resistance is defined as the pressure drop in a segment caused by a fluid (gas) flow through it, and divided by the magnitude of the same flow. This resistance is determined by the frictional losses, laminar or turbulent, in the moving fluid (gas).

Breathing resistance is usually defined more generally, by saying it is the total effort the respiratory muscles must produce to generate a certain respiratory ventilation, divided by this ventilation. Perhaps it would be more correct to divide by the *useful* ventilation, that is to exclude anatomical or apparatus deadspace ventilation. Both deadspace ventilations may be changed by the breathing apparatus design. This definition is not usual and will not be used here either, but should be kept in mind. If the effort of the respiratory muscles needs to be increased significantly, this would certainly not be felt as comfortable. On the other hand factors not affecting this effort could *also* make the respiration feel uncomfortable. Examples of this are: necessary effort by other muscles like cheek and mouth, conditioning of the breathing gas like temperature and humidity, which could at the same time also affect the respiratory resistance[5] and contaminants (smell).

The effort required by the respiratory muscles can be given by the work they are producing. This work of breathing (WOB) is determined by the pressure required to make a change of lung volume, integrated with the volume change.[6] The pressure is determined by elastic, resistive and inertial components, which make elastic work, resistive work and inertial work. The latter, indicating what is necessary to accelerate and decelerate gas and tissue is normally cancelled out over a breathing cycle and usually ignored. The elastic work is determined by the elastic properties of the lungs and breathing apparatus (for example by a breathing bag). It is important that elastic properties of the lung are different at different lung volumes. Because of this the elastic work will increase if a subject is forced to breathe at a different lung volume than normal. This is what happens when the pressure at the lung level is increased relatively to the mouth pressure (HI) as in vertical immersion. The flow resistant component is determined by the frictional losses. Some of this friction is made by the tissue, but normally we only consider the friction in the gas flow through the human airways and external apparatus. This friction in the gas stream is determined by the geometry of the flow segments, the magnitude of the flow and the properties of the gas. It is increased by narrow passages, sharp edges, increased flow and the density of the gas. The total WOB can approximately be measured by the pressure in a balloon inserted in the oesophagus at the end of a tube, when at the same time lung volume is measured.

The total WOB is divided into external and internal work. The external WOB is defined as what can be measured at the mouth level and the internal is the remaining, i.e. what is done in the body. This separation is mainly done for practical reasons. The external WOB is much easier to measure and the conditions that affect it are easier to define since specific physiological characteristics for lungs and airways do not have to be considered. Also once you have determined ventilation and breathing pattern the external WOB is a specific value for the equipment used under the specific conditions. But it is important to keep in mind that it is the total WOB that is important for the diver and this will not only be a function of the external WOB, but also, apart from individual physiological characteristics, density of breathing gas, reference pressure (HI), relation of inhalation to exhalation pressures and breathing pattern.

The total WOB is important because it, together with the strength of the diver's respiratory muscles, determines the magnitude of respiratory ventilation that can be maintained for a length of time and also the maximum breathing capacity. It is well known that the ventilation increases the proportion to the amount of physical work a person is doing. The increase in ventilation may be less in a diving situation with increased WOB, but this will induce a carbon dioxide retention which can be tolerated to only a certain level. So if the ventilation is not increased, the work rate must be limited. It is generally accepted that while physical work capacity in the normal, 1 atm dry situation is not limited by the breathing capacity, it will be the limiting factor in deep diving. Thus the total WOB will, together with the endurance of the respiratory muscles, determine a maximal work endurance, and together with the strength of respiratory muscles put a limt on the short-term physical work capacity.

The external WOB will after what is said above, always be just a part of the total WOB. An example of how the relationship could be is shown in Fig. 1. The data are obtained in a simulated dive at 400 msw with a diver exercising on a bicycle ergometer. The upper collection of points (circles) shows how the total WOB is increasing as ventilation is increasing, while the triangles beneath indicate the external WOB. It is seen that the external part of WOB is only 20–30% of the total. It is clear that even if the breathing apparatus could be designed without external WOB at all, the diver will still have a problem with the internal WOB that is increased because of increased density.

It would be most correct to measure the total WOB on a number of divers using the breathing equipment under realistic conditions, to evaluate the breathing resistance. In pratice this would be too expensive to do in evaluation of every equipment. Generally it is assumed that by measuring the external WOB plus peak pressures and HI, in an unmanned test on a breathing

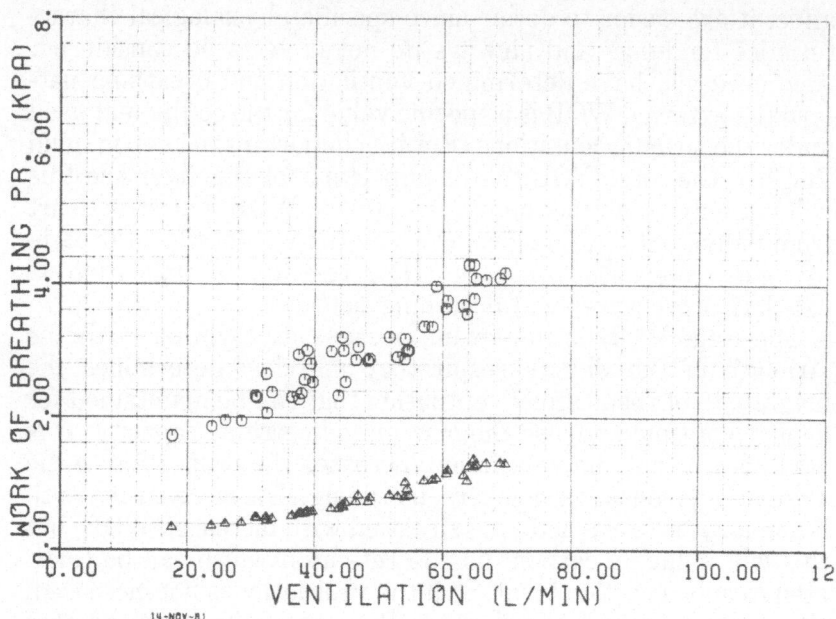

Fig. 7.1 WOB plotted as function of ventilation for a diver working on a bicycle while breathing from a modified[1] UBA at 400 msw simulated depth. Circles are total WOB, measured by a oesophagal pressure transducer, triangles external WOB measured by a pressure dransducer at mouth level

simulator at standardized conditions it is possible to predict what the total breathing resistance will be. On this assumption it is possible to set limits on external WOB, HI and maximum mouth pressure fluctuations and in this way be sure that the total WOB or breathing resistance is not excessive. This is probably nearly true, but only nearly. There is, for example, not enough information to exactly predict how a diver will adjust his breathing when exposed to different external breathing resistances in various conditions. Also the internal WOB will increase with increased density of the breathing gas regardless of how the external WOB varies with depth. We can, for example, assume an imagined breathing equipment that has the same characteristics at different depths. That these characteristics were acceptable by the diver at a shallow depth would not mean that the same was true at a greater depth. At a greater depth the internal WOB could have increased so much because of the increased gas density that it would not be possible for the diver

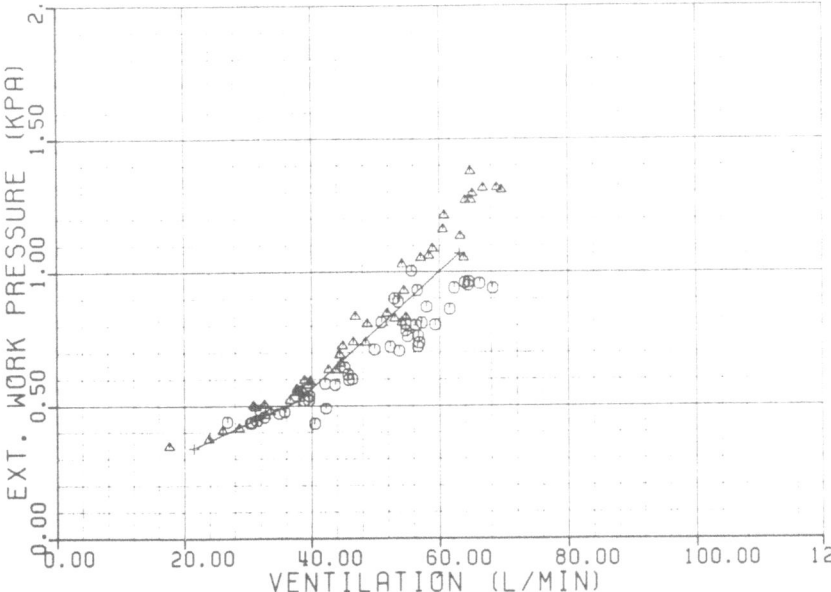

Fig. 7.2   External WOB as function of ventilation for one unmanned test (crosses connected by lines) and two manned tests (triangles for one subject, circles for the other). The UBA is the same as in Fig. 1 and a simulated depth 400 msw is also used here

to stay up with the total WOB. Normally a sinusodial breathing pattern is used on unmanned tests. This is only done for simplicity; a diver will never breath like this. How the diver adjusts his breathing rate and pattern will affect not only the internal WOB, but also the external. This is shown in Fig. 2.

   The points for the manned tests have a certain spread around the line indicating the unmanned test results. This could mean that the breathing resistance of the equipment changed while the test was going on, but this is not thought to be the reason. It is more likely that it was caused by differences in breathing rate and shape. If this is true it can also be seen that one diver tended to use the equipment more efficient than the other.

REQUIREMENTS FOR BREATHING RESISTANCE

The ideal UBA shoud not only add no external breathing resistance to respiration, but should also assist ventilation so the increase of internal resistance (WOB) caused by increased gas density is counteracted. If possible, assisted ventilation would be more and more advantageous at greater depths. If we are limiting the possible depth to about 400 msw for heliox diving, this ultimate, ideal requirement is probably not necessary, but is nevertheless something to aim for.

If we, instead of this 'ideal' requirement, try to put up a 'realistic' requirement, we must first say what we mean by 'realistic'. We want to define 'realistic' requirements for breathing resistance so that total cost to fulfil them is justified by total gain in safety and efficiency. Following this definition, it is clear that other requirements important for safe and efficient diving must be compared to the breathing resistance requirement. Examples of such other requirements are:

- a simple design, fail safe and easy to maintain;
- not too heavy, does not restrict movement, is not easily entangled;
- does not restrict vision or impose too much noise;
- breathing gas conditioned with respect to pO_2, pCO_2, temperature and humidity;
- possibility for emergency life support.

The situation today is that all those other requirements imply that the realistic requirements must be set less stringently than we could wish if they were ignored. It would, for example, be no good to have a UBA with super breathing resistance characteristics if it was so heavy that the diver could not lift it or move around in the water with it. Neither would it make any sense to reduce the breathing resistance if this caused the noise level to be so high that the diver could not communicate.

On the basis it should be clear that selection of 'realistic' requirements will be affected by feelings and subjective opinions.

As explained above, it is practical to base the requirements on parameters that can be measured in an unmanned test. While this is so we must again stress that it is the *total* WOB that is important and that this is not uniquely determined by the external parameters, especially not with increasing depth. In addition to requirements for external WOB, requirements for HI and maximal pressure fluctuations at the mouth will help in defining the total WOB requirements. HI and maximal pressure fluctuations will, in addition, have implications for the feeling of comfort that is not connected to total WOB.

In our opinion Morrison's proposed performance goals are the best existing basis for realistic requirements for breathing resistance, but the NEDU goals are also useful. Both state that requirements should be made for respiratory ventilations up to 75 litres per minute, which is in accordance with our experience and opinion. On the HI the two proposals use different definitions. Morrison is comparing reference pressure against the lung centroid, which is the right place, if total WOB is concerned. NEDU is using the suprasternal notch instead of lung centroid, taking into consideration that the feeling of comfort is affected by pressure differences at the mouth level as well. The big problem with requirements for HI is that although it is clear that the ideal reference position is not the mouth level, but should be closer to the suprasternal notch, it is not realistic to set the requirement such that the demand regulators with a reference diaphragm by the mouth are all excluded yet. For maximum mouth pressure fluctuations and external WOB NEDU's goals are quite similar to Morrison's 'recommended' levels. Morrison's 'acceptance' levels are significantly less stringent than these two. NEDU does not give 'acceptance' levels, on the contrary they state that: 'these goals do not represent minimum acceptable performance levels'. Morrison defines 'recommended' as level of comfort and 'acceptance' as level of tolerance.

NEDU goals for mouth pressure fluctuations are for much equipment harder to meet than the external WOB figures. In our opinion this is not a 'realistic' proportion for these two parameters, because external WOB must be recognized as the most important, and also the peak pressure is less well defined. We are *not* saying that the figures are wrong, based on NEDU's definition of those goals, *only* that they are not suitable as 'realistic' requirements following our definition. The same could perhaps be said for Morrison's peak respiratory pressures, but certainly to a smaller degree. What is at least certainly needed is a strict definition of what is meant by 'peak pressure' (for how long a time must a peak last to be defined as a 'peak' and not noise). With this lacking, the speed of the recorder could determine if equipment is passing the requirement or not.

For external WOB Morrison's 'recommended' levels are less sringent than NEDU goals on low respiratory ventilations, on the highest, 75 litres per minute, he has proposed a slightly stricter value. We could say that he is more concerned with the emergency situation and less with the working situation compared to NEDU.

The most important reason to give both 'acceptance' and 'recommended' levels was that recommended levels were not within reach of the state-of-the-art for UBA development. If all diving should not be stopped immediately, recommended levels could not be set as requirements immediately. 'Acceptance' levels were seen as a compromise between what was possible and desirable and at the same time tolerable. It is fair to say that in 1981 the proposed acceptance levels could be used as what we have defined as 'realistic' requirements. This reflects both the general status of UBA development at that time and the operational depths in question (up to around 200 msw). Since then the UBA development has improved and the possible operational depth has increased, which means that total breathing resistance (WOB) will be bigger even if the same levels of external WOB are followed. To keep requirements in general at the 'old' acceptance level would, in our opinion, be to keep them 'unrealistically' high. In some cases, however, these acceptance

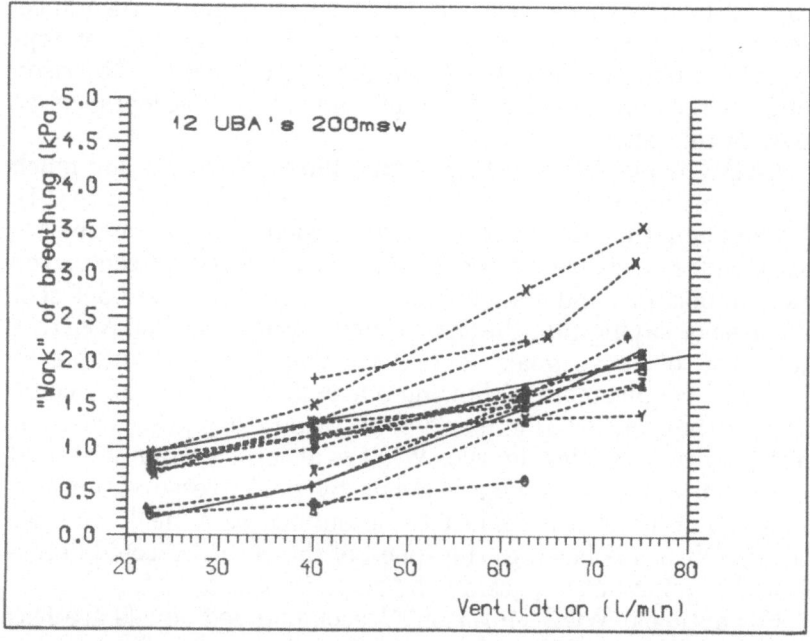

Fig. 7.3 Unmanned test results for selected breathing equipment (UBAs). Solid broken lines show NEDU[4] category 5 goals and solid straight line Morrison's recommended levels[1]

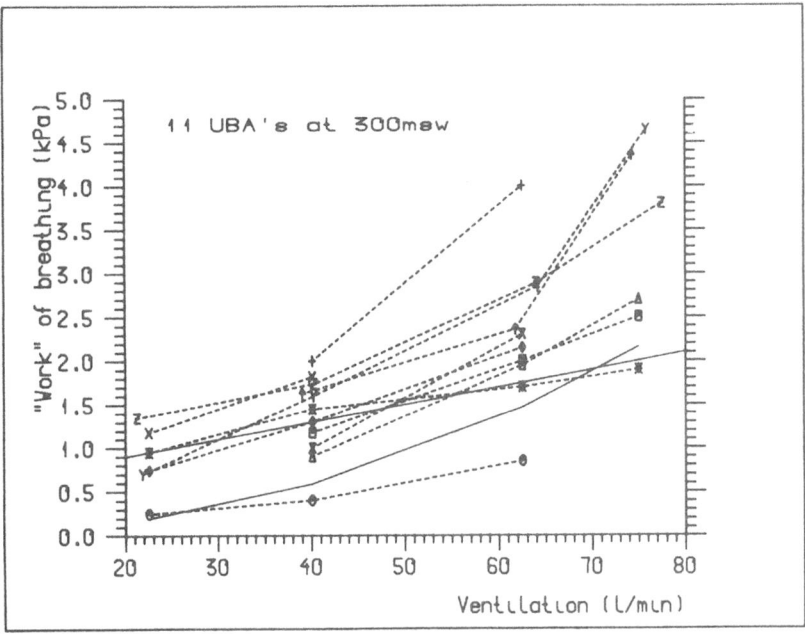

Fig. 7.4 Unmanned test results for selected equipments (UBAs). Solid broken lines show NEDU[4] category 5 goals and solid straight line Morrison's recommended levels[1]

levels can be kept and considered 'realistic';

- It is impossible to have equipment with less resistance and at the same time keep other requirements important for safety and efficiency.
- The maximal operational depth is shallow (not much more than 200 msw).
- Emergency situations that require hard physical work are very unlikely to occur.

We think that 'realistic' requirements today are closer to Morrison's recommended levels for external WOB. The reason for this can be found in Figs 3, 4 and 5, which summarize some of the unmanned UBA-tests that have been done at NUTEC over the last few years. The results are compared with recommended levels and NEDU's goals (the strictest category). Not all equipment has been measured at all required respiratory ventilations. But it can be seen that 6 to 7 UBAs are within

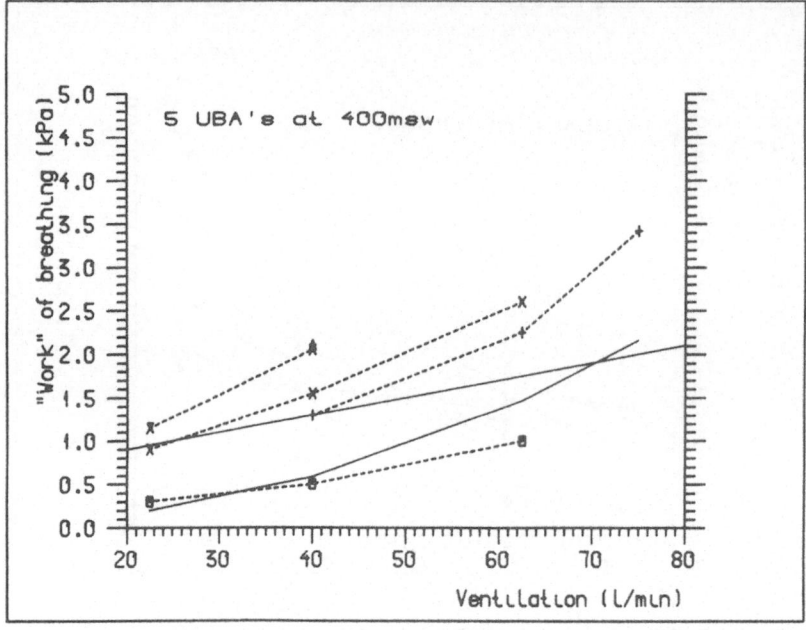

Fig. 7.5 Unmanned test results for selected breathing equipments
(UBAs). Solid broken lines show NEDU[4] category 5 goals and solid
straight line Morrison's recommended levels[1]

recommended levels at 200 msw. One UBA is within at 300 msw
while one is almost within recommended levels here. At
400 msw only one UBA is within recommended levels.

We think that the recommended level, which would allow a
number of the items of equipment to be used at 200 msw, only
very few at 300 msw and 'almost none' at 400 msw, used as
'realistic' requirements would reflect other aspects of diving
technology of today, in a correct way. In other words; diving is
today definitely possible in a safe manner to 200 msw, probably
possible to 300 msw but very hard to accomplish to 400 msw.
Compared to the NEDU goals only one UBA may be good
enough at 200 msw (dependent on the missing value at 22.5
litres per minute), while another is very close at all depths. This
does not reflect the present status of diving techniques and is
not suitable as 'realistic' requirements, but may be all right as
'goals' which is what they are defined as. It is our hope that in a
few years these limits which today are goals could be defined as
'realistic' requirements.

What we have said above must not be interpreted that we mean governmental regulation should apply to the recommended levels. We have specified indications justifying violations and thus they can not serve as absolute regulations.

The acceptance levels, or maybe slightly stricter, would in our opinion be suitable. The authorities and operators should, on the other hand, not consider this to be enough for all circumstances; especially for great depths and dangerous conditions recommend levels should be indicated. Otherwise the acceptance levels would become a brake on the development of breathing equipment.

CONCLUSIONS

Breathing resistance can be considered as roughly identical to total work of breathing (WOB). This is a very important factor in limiting the physical work capacity of a diver.

Total WOB is not solely an image of external WOB, but is also importantly affected by gas density, hydrostatic imbalance, breathing pattern and technique as well as individual physiological parameters.

'Realistic' requirements for breathing resistance must be defined with reference to other requirements affecting safety and efficiency.

In our opinion Morrison's recommneded levels for external WOB are suitable as 'realistic' requirements, acceptance levels should serve as absolute requirements while NEDU's goals for the lower ventilations are too strict to serve as 'realistic' requirements.

REFERENCES

1. J. B. Morrison, 'Physiological acceptance criteria for underwater breathing apparatus' and 'Unmanned test procedures for underwater breathing apparatus' prepared for Petroleum Production Division, Department of Energy, 1981.
2. B. H. Hjertager and T. Nome, 'Conclusions from the chairmen'. Workshop on diver's breathing equipment, Bergen 30 Nov. to 1 Dec. 1981.
3. D. H. Robertson, 'Guidance notes on underwater breathing apparatus'. Offshore Research Focus No. 45, Oct. 1984.
4. J. R. Middleton and E. D. Thalman, 'Standardized NEDU

   unmanned UBA test procedures and performance goals'. NEDU
   Reprt No. 3–81.
5. I. Rønnestad, 'Humidified breathing gas'. Thesis, NUTEC, 1984.
6. J. B. Morrison and S. D. Reimers, 'Design principles of
   underwater breathing apparatus'. In *The physiology and medicine
   of diving*, (Ed P. B. Bennet and D. H. Elliot), 1982.
7. B. Schenk, D. Furevik, K. Segadal and R. E. Peterson, 'Final
   report on Deep Mask 81 Program'. NUTEC Report No. 10b–82.

8

Increasing Bottom Working Time: Reducing the Decompression Penalty

C. M. Childs, Occupational Health Ltd., Aberdeen

INTRODUCTION

If the subject of this paper were simple and easy to resolve, it would not have been included in this proceedings. Increasing effective bottom working time, while keeping the decompression penalty as small as possible, is perhaps the central aim of all diving techniques. It is for this reason that the various diving techniques have been developed but only with saturation diving has the decompression penalty for unlimited working time been reduced to zero, at least until a series of working periods has been completed.

REVIEW

It would be useful at this stage to review the working times available using examples of different diving techniques, and the factors that act as constraints upon increasing working time and decreasing the decompression penalty (Table 1). Fundamentally, with all diving techniques, the limiting factor is the amount of inert gas taken up by the diver during the period under pressure and the time it takes during decompression for this to leave the diver. An allied limiting factor is that of oxygen toxicity. Pure oxygen cannot be used as a breathing mixture,

TABLE 1
Limitations on uses of diving techniques

	Scuba diving	Surface supplied diving	Bell bounce diving	Saturation diving
Breathing air	165 fsw	165 fsw	165 fsw	165 fsw
Breathing heliox	Not used in UK but not illegal. No commercial requirement	To 300 fsw. Not usually used; requires in-water O_2	To 300–400 fsw. Decompression increases with time	? max. depth. Only one decompression per working series

except in highly specialized military applications, because of its toxicity. This limits its use in replacing part of the inert gas of breathing mixtures in attempts to reduce inert gas uptake.

Considering scuba first, its use is limited to 165 feet in commercial diving because of the risk of nitrogen narcosis. In practice there are other potential problems with the technique such as in-water decompression stops, which make it less than ideal for routine use in the commercial diving industry. For example the maximum bottom time requiring no in-water stops for a 130 foot air dive is around 10 minutes, and for 165 feet it is around 5 minutes. Surface supplied air diving is, in effect, the diving industry's equivalent of scuba diving. Thus the constraints in use are similar, and there is the same depth limitation of 165 feet. Deeper than this a different breathing mixture must be used. The same limitation is imposed by nitrogen narcosis as during scuba diving, although the potential exists for an alternative heliox breathing mixture. This is generally not popular because of the extra cost and because in-water oxygen breathing is regarded as unsafe diving practice.

Moving to the next diving technique, that of bell bounce diving, there is little theoretical limitation to its use as the depth limitation imposed by nitrogen narcosis can readily be avoided by using a heliox breathing mixture. In practice, however, for dives of more than about 350 feet, with a bottom time of more than 60 minutes, most diving contractors would probably regard saturation as a more suitable technique because of the long decompression required for bell bounce diving. Due to the extremely limited bottom times possible during deep bell bounce dives it is necessary to pressurize the divers as quickly as possible to try to maximize the bottom time. A rapid compression to depths in excess of 400 feet might possibly produce symptoms of the high pressure nervous syndrome. A diver suffering from these may take some time to orientate himself after leaving a diving bell, reducing the effective bottom time. Typically, for a 60 minute dive to about 350 feet, 32 hours are required to decompress a diver. Unless additional chamber facilities are available, another dive could not take place until the previous divers have left the chamber. If the divers failed to accomplish their task for some reason while on the bottom a delay would be incurred. Using a team of 6 divers in a saturation mode, however, virtual round-the-clock diving operations can be conducted with each pair of divers being able to spend 8 hours in the water. An added advantage of saturation diving is that the incidence of decompression sickness is far greater during a bounce decompression than during a suturation decompression.

WORKING PRACTICES

Before returning to the physiological determinants of bottom working time and decompression penalties, there are practical areas which need to be mentioned. These include dive planning, diver equipment, and the design of divers' tasks. To maximize useful working time the supervisor must ensure that the diver is positioned as close to the work site as possible to reduce the distance a diver has to move through the water, wasting available working time. To do this the supervisor must ensure that the cage or bell is lowered so that it is close to the work site and so that the effects of any currents against the diver are minimized. The efficient use of the available working time can also be maximized by familiarization with the task and by training. Proper design and maintenance of diver equipment will also make for the efficient use of available time, for example, the use of specially designed or modified tools, and thoughtful design of the diver's personal equipment so that he is as little encumbered as possible and as free as possible to carry out his work. Complete failure of any equipment will clearly lose working time, but partial failure will also waste time. The design of the diver's tasks will also contribute to the efficient use of available working time, and this must begin during the design and construction stage of any equipment or structure on which a diver may be required to work.

The use of enclosed bells and transfer under pressure in all air diving would very effectively add to a diver's working time as he would not have to leave the worksite to begin his ascent. There would be, however, a considerable penalty in equipment costs if this practice were generally accepted, and could only be considered if operators were convinced of the value of this service.

PHYSIOLOGICAL DETERMINANTS

Earlier I summarized the factors limiting bottom working time as being the amount of inert gas taken up by the diver and the time taken for this to be lost. These two factors determine the decompression penalty and thus the decompression rate. Examining the two factors in turn will demonstrate the considerations which have been given to prolonging working time and decreasing the decompression penalty by reducing inert gas uptake and increasing its elimination.

First, it is possible to reduce the inert gas component of a breathing mixture by increasing the oxygen partial pressure. There are clear limitations on this in the short term because of central nervous system oxygen toxicity and in the longer term because of pulmonary oxygen toxicity. Although the signs and symptoms of these are well known, there is still no reliable means of predicting their occurrence. Thus most diving procedures work within a range of oxygen partial pressure which is thought to be safe and indeed in many cases may be unduly conservative. This is particularly true in shallow air diving where, with a maximum worksite depth of 30 metres, Comex Houlder Diving opted for the use of air diving techniques using a systematically planned nitrox system. The use of nitrox allowed the work period to be increased by approximately 66 per cent, without decompression.[1] This type of enhanced oxygen breathing could be routinely carried out by commercial diving contractors and demonstrates one approach to increasing bottom working time. The technique of using a nitrox breathing mixture containing less nitrogen than air and more oxygen has a well established history in military diving, where it is given the title of 'equivalent air depth'. It is based on the fact that, when using oxy-nitrogen mixtures in closed-circuit equipment, the percentage of nitrogen in the mixture being breathed is considerably lower than that in air. In consequence, by Henry's Law, the amount of nitrogen absorbed by the body at any given depth on the mixture would be considerably less than would be absorbed if air were used at the same depth. If standard air decompression procedures were used for such a dive, the time spent on stops would be unnecessarily long. Using an established calculation it is possible to calculate, from the percentage of nitrogen breathed, the equivalent air depth of the dive undertaken and to decompress accordingly. For any depth used in the air range, therefore, an extension is possible to the bottom time by decreasing the amount of nitrogen in the breathing mixture by increasing the amount of oxygen. The technique is, of course, limited by the constraints of oxygen toxicity and the amount of oxygen that can be substituted for nitrogen will be limited. One of the recommended mixtures suggested for military diving contains 60% nitrogen and 40% oxygen.

A second approach is to increase the rate of loss of inert gas from the diver during decompression. A report from Duke University illustrates such an attempt during saturation decompression from deep dive trials. Although the report is of attempts to find a suitable decompression table rather than to reduce decompression time, the problem is the same as it would be in an attempt to reduce the decompression penalty and thus maximize bottom working time for a given decompression period. During decompression from 1000 feet the divers breathed trimix from 1000 feet to 850 feet, with a partial pressure of oxygen of 0.8 (Table 2). No decompression sickness was reported but symptoms of pulmonary decompression sickness developed in a number of divers. In the next series of dives the partial pressure of oxygen was reduced from 0.8 to 0.6 and there were no symptoms of pulmonary oxygen toxicity. The penalty

TABLE 2
Duke University decompression trials

Decompression from 1000 fsw to 850 fsw		
	Series 1	Series 2
pO$_2$	0.8 bar	0.6 bar
Decompression time	– – – – – – – – –	25% greater

for this decrease in pO_2 was that the length of decompression was increased by 25% because a slower ascent rate had to be used. Reversing this argument, decompression times can be reduced by increasing the oxygen component of the breathing mixture, thus increasing inert gas elimination, but the procedure is limited by the constraints of oxygen toxicity. Attempts were made in the 1960s by Buhlmann and Keller to increase inert gas elimination during decompression without incurring penalties from oxygen toxicity. During decompression they switched from one breathing mixture of inert gas to another to encourage inert gas elimination and were able to undertake successful dives to depths which were, at the time, considered impossible. Although the success of their work was marred by a fatal accident the technique was successful and is again attracting interest.

A third approach is to use inert gases other than pure heliox during compression to improve working ability when the diver reaches his working depth during a saturation dive, and to use other inert gases during the time at working depth to improve efficiency by increasing thermal and respiratory efficiency. In a long series of dives Bennett at Duke University has investigated a trimix of nitrogen, helium and oxygen to reduce the effects of the high pressure nervous syndrome. These would otherwise make divers at best inefficient and at worst incapable following a rapid compression. The evidence for the favourable effects of trimix are still debated, however, with claims being made that trimix only masks the effects of the high pressure nervous syndrome by the narcotic effect of nitrogen. There is no doubt, however, that increased working efficiency of dives to great depths could be increased by avoiding the recovery period which has to be included on heliox dives for the divers to overcome the effects of a rapid compression. Other gas mixtures have been investigated for use at depth, with the intention of increasing diver efficiency and work capability by improving voice performance, thermal efficiency and the work required for breathing. The principal mixtures that have been tried include neon or hydrogen as partial replacements for helium. It should be possible to improve voice performance and thermal difficulties by technical advances, but respiratory work, largely determined by gas density and thus breathing resistance, is at present one of the greatest limitations on useful work being performed at great depths. Hamilton and Kenyon[2] reported a series of dives using a mixture of neon and oxygen in which they successfully reached 640 feet, but in which the divers still experienced breathing difficulties using standard Kirby-Morgan breathing equipment. Recently Comex have undertaken a series of dives using a mixture of hydrogen, helium and oxygen in which they reported advantages in breathing resistance at about 400 feet, although they found the narcotic effects of hydrogen to be significant and a possible limitation.

SUMMARY

There are several ways in which the efficiency of use of available bottom working time can be increased. Careful planning of the dive by the supervisor, involving attention to the diver's position relative to the worksite and the manoeuvring necessary for him to be ready to start work, can increase efficiency as can task training and appropriately designed personal equipment and tools. Anticipation during the design and construction stages of equipment and structures divers may have to work on should reduce the time it will take for work to be done.

There is still scope for study of the alteration of the inert gases used during compression, stable depth and decompression with the aim of improving the diver's ability to work on reaching working depth, his working efficiency at depth, which is in large part determined by breathing resistance, and increasing inert gas elimination during decompression by inert gas switching.

The extension of bottom working time and reduction of the decompression penalty by alteration of the amount of oxygen in a breathing mixture during a dive and during decompression can be effective, but is rather more difficult to achieve safely. The difficulty stems from our lack of knowledge about oxygen tolerance limits, both at steady working depths, when it might be used to reduce the inert gas uptake by the diver, and during decompression, when a higher partial pressure of oxygen should encourage the elimination of inert gas. In addition to the problems of oxygen toxicity there is also the suspicion that oxygen can actually contribute to the size of the bubbles of decompression sickness if breathed at relatively high partial pressures during decompression. This has been reported in the past both from practical experience and from theoretical analysis of the movement of both inert gas and oxygen within the tissues.

There is still a requirement for further work on the effects of oxygen in its toxic action on the lung and central nervous system and in the effectiveness of a higher respired pO_2 in allowing increased inert gas loss from the body during decompression. In the context of further studies of oxygen toxicity a useful tool both for this research and for operational diving would be a predictive system to replace the UPTD system devised by Lambertsen. This system was not designed for the uses to which it is now put, but we have no alternative. Such an alternative would make easier the study of increasing bottom working time without incurring a greater decompression penalty.

CONCLUSION

There are ways in which the efficiency of any working dive can be improved by relatively simple means such as planning and design. Other methods, aimed at counteracting or avoiding the effects on physiology under pressure are more complex and in many ways still uncertain. Oxygen toxicity and the advisability and usefulness of different inert gases would undoubtedly merit further study.

REFERENCES

1. Comex Houlder Diving Ltd., *Information Bulletin* No. 18.
2. R. W. Hamilton and D. J. Kenyon, *Decompression work at Tarrytown, Development of Decompression Procedures for Depths in Excess of 400 feet*, Undersea Medical Society, 1975, p. 95.

9

Increasing Bottom Working Time: Improved Methods of Diver Deployment

Derek Clarke, Managing Director of Mara
Engineering Ltd

INTRODUCTION

This paper reviews the present means used to deploy the diver
closer to the job site and thereby increasing bottom working
time. The paper does not address the situation where the support vessel can locate directly over the work area.

The six basic means adopted at present are as follows:

(a) Cross haul of diving bell towards job site.
(b) Deploy diving bell/basket from a boom.
(c) Guide wire system to guide bell to job site.
(d) Mobile diving bell.
(e) Diver lock out submarine.
(f) Orientate diving from above the job site, i.e. platform
based.

The increased use of large semi-submersible vessels has stimulated development in this area simply due to the stand off
required and this is particularly apparent when supporting surface diving on an adjacent structure.

CROSS HAUL OPERATION

The technique of cross hauling is well established and has been used on diving support vessels with moonpool launched bells for a number of years. The method is to attach a second lifting line to the diving bell, via a monkey plate, which passes down through the moonpool and back up to the side of the vessel. This wire passes over a guide sheave or fairlead to an independent winch.

After launch, the weight of the bell is transferred to the cross haul line, which moves the bell closer to the side of the vessel. The amount of movement can be increased with the aid of extension booms.

This technique has also been used on drill rigs and platforms, and in a developed form on the BP Iolair and Shell Stadive. Both vessels, being large semi-submersibles, are at a disadvantage when compared to a smaller monohull DSV for working alongside a platform. Both vessels have long booms to improve outreach and employ motion compensation on the cross haul line.

Cross hauling is a simple and cost effective means of moving the bell but there are inherent disadvantages:

(a) Limited lateral movement — cannot reach cross hauling point due to catenary tension in main lift wire.
(b) Difficult to monitor exact position of the bell.
(c) Cross haul wire is subject to corrosion and wear and must be frequently inspected.
(d) When cross hauling, conventional guide wires are not used which maintain orientation of the bell. This can give problems with the recovery of side mate bells into cursors in particular.

BOOM DEPLOYMENT

Boom deployment is different to cross hauling in that the bell/ basket lifting wire actually passes over the end of the boom. An adaption of this is to 'push' the lifting wire further outboard after the bell has been deployed. Such a system is used effectively on the Philips SS, a semi-submersible diving support vessel, when operating alongside a platform, with the saturation bell.

Cranes can be considered as a form of deployment boom and

these have been used to lift dry diving platforms from which the diver makes free descents to the job site. The diving platform would house the diving supervisor, standby diver and tenders and would be provided with autonomous supplies.

Booms must be robustly designed and mechanically very reliable. A situation resulting in the boom jamming in the outboard position would be hazardous to a diver requiring decompression, if being deployed by a wet bell or basket.

To overcome this potential hazard, one operator is considering the fitment of a boom which incorporates a walkway. Diving would be carried out directly from the end of the boom using conventional wet bell/basket, with the DDC being located on the deck. Notwithstanding the structural constraints this concept is limited by the safe distance a diver can travel in the restricted 'surface interval' associated with surface diving.

GUIDE WIRE SYSTEM

This system has been used on drill rigs and fixed platforms. The guide wires are connected adjacent to the job site and the bell/basket runs down the wires. This method is limited and generally the lateral distance moved will be up to 25% of the vertical drop. If applicable it is a very simple method; however, the wires must be inspected frequently.

MOBILE DIVING BELL

The concept of a diving bell that can move under its own power has been around for some time. A number of designs have been put forward, however there is only one working system, Occidental's Tharos. The MDU (Mobile Diving Unit), built by Perry, is a two compartment vehicle with a pilot capsule mounted above a lockout capsule. The unit has logged over 790 dives.

The MDU has proven to be an effective machine for getting the diver to the job site, however one of the principal advantages is the ability to access a different side of the jacket without moving the vessel or alternatively, being able to remain at the job site during periods when the vessel is required to locate some distance from the platform. This would be a particularly useful facility if the flare position prevents access to one face. Any operation involving a mobile bell requires careful planning to consider the umbilical. With tidal changes, snagging of the adjacent structure can be a hazard.

The excursion radius when supporting can typically be up to 250 ft; however, on trial without latching on to the worksite it has excurted over 1000 ft horizontally at a depth of 360 ft and hovered in a stationary position for over 30 minutes.

Stena's Seawell vessels will have a mobile (flying) bell in addition to the conventional bell. The mobile bell will be piloted to the job from the surface in a similar manner to an ROV. Final positioning and latching will be under the control of the diver from within the bell.

The mobile bell will be deployed from a catcher frame which can be lowered to a depth of 50 to 100 m as required, prior to the excursion.

The Stena approach results in a smaller unit requiring less power and a smaller umbilical when compared to Oxy's MDU, however, at the expense of a more complex diving bell.

LOCK OUT SUBMARINE

In the mid-seventies mini-submarines fitted with a pressurizable divers' compartment, to support autonomous lock outs were heralded as the new way ahead. A number of contractors responded to this prediction but the principle foundered due to lack of power and gas storage and the inability to support the diver from the surface. In spite of developments in self contained powered sources (closed cycle diesel, Sterling engine and small nuclear plant) closed circuit breathing systems and high energy density heating systems (such as exothermic chemical reaction, hydrocarbon fueled burners and latent heat dissipation from solidifying hydrates) the momentum has been lost.

These developments can provide life support for the divers over a similar period to a saturation bell run; however, the main attraction, autonomy from the surface, is also the main disadvantage when the diver requires additional tools, materials or specialized equipment.

In situations where surface support is neither desired nor possible, such as under ice, then the submarine approach is attractive. Presently projects by Comex, Kochums, Sub Sea Oil Services and Bruker are incorporating the above developments into larger mini-subs able to support autonomous operations over a number of days with a range of several hundred kilometres.

PLATFORM/RIG BASED EQUIPMENT

A major problem when supporting a diver from a large vessel lying alongside a platform is the lateral excursion that might be made. This requires long umbilicals or the use of one of the methods described above, to position the bell closer to the job site.

The siting of the support equipment on the platform or floating installation has been frequently employed on both economy and operational grounds. Leaving the financial justification to one side, the deployment of divers from the installations offer the following advantages:

(a) Increased weather window due to stable platform.
(b) Diver can be deployed close to job site perhaps assisted by guide wires or cross hauling.
(c) Increased safety.
(d) Support facilities (e.g. burning, water jetting, rigging) can be provided directly over the job site.

There are, however, some disadvantage:

(a) Weight and space consumption on platform.
(b) Increased maintenance task for platform.
(c) Evacuation by divers under pressure during a platform emergency could be impossible.

Diving facilities have been temporarily installed to support a heavy but specific diving programme but there are numerous installations which are permanent. It is interesting to note that the majority have been retrofits. Table 1 lists those known to the author.

In the table, 'fixed or mobile' refers to the diving facilities. In some cases it has been necessary to make some or all of the diving facilities mobile in order to give diver access to the whole structure.

In briefly analysing this table one can see that the majority are afterthoughts, justified principally on economic grounds due to a high level of unplanned diving activity. The majority are to

TABLE 1
Diving facilities available on N. Sea platforms

Installation name	System type	Retrofit	Fixed or mobile	Comment
Piper	air	yes	fixed	
Claymore	air	yes	fixed	
Heather	air	yes	fixed	
Ninian Central	sat	yes	mobile	
CDP-1	air	yes	mobile	
Magnus	air	no	fixed	being commissioned
Brae A	air	no	fixed	being commissioned
Brae B	air	no	fixed	under design
MCP-01	air	yes	mobile	
Argyll	sat	no	fixed	contractor's plant
Murchison	air	yes	fixed	contractor's plant
Hutton TLP	air	yes	fixed	contractor's plant
Buchan	sat	no	fixed	
Balmoral	sat	no	fixed	under constrution

support air diving as there is a high proportion of work done in this range and due to the relative simplicity and lightweight nature of the equipment.

THE FUTURE

Predictably, the simple methods for moving a bell close to the job site will stay with us as long as diving does.

The number of mobile bells will increase and the application of ROV technology to remotely pilot the bell to the job site, whilst acoustically monitoring its position, is the direction these developments will go.

The increased acceptance of the mobile bell has stimulated some interest in adaption kits to convert standard bells to have buoyancy and thruster facilities.

It is also conceivable that a pressurizable air bell, fitted with a thruster/buoyancy control assembly could be used to support air diving. Such a device would not commit the diver to going under pressure until latched on to the job site. The operation would be treated as a bounce dive, with decompression carried out both in the bell and in a conventional Deck Decompression Chamber. This procedure would maximize on working time for the minimum decompression. Such a device could be made economically since it is only working at low pressures and requires minimal autonomous life support.

With platform inspection, maintenance and remedial works ever increasing during the life of the structure, the air diver is assured of an ongoing demand for a long time. Such a device could be deployed from a vessel or platform. The argument that air diving does not require sophisticated equipment is erroneous when one considers that a vessel costing between £15000 and £25000 per day could be engaged in doing nothing but supporting an air diver. It is with air diving that there is the greatest scope for increasing bottom time by using radically improved methods of deployment.

10

ADS and ROV Systems in Support of Divers

J. Balch and P. Sheader, OSEL Group Ltd

INTRODUCTION

This paper is a general discussion of the supportive role available to divers from ROV and Atmospheric Diving Systems (ADS).

It is hoped that the paper will show that developments in submersible technology can be considered as complementary to diving programmes and not as is generally thought, totally competitive.

HISTORICAL PERFORMANCE

Vehicle support of divers has in the past been limited mainly to the use of the 'eyeball' ROV, primarily as a general surveillance unit to aid surface controllers' engineering and supervisory input to the divers' activities.

This developed to diver assistance in terms of lighting, video of pertinent aspects to the surface, navigation to some extent, and the ability to relieve divers in some circumstances of the need to carry camera units.

Many well documented cases are available which have highlighted the increase in general diver safety, and efficiency in reducing dive time, due to the increased data available on the surface.

Divers over the last few years have accepted this aspect of ROV technology to a point where the eyeball ROV is considered to be a valuable aid and tool a standard part of a diving programme.

Indeed, in some recent diving programmes, divers have questioned the absence of the eyeball from their dives.

General structural inspection by ROV is usually considered as a separate activity or operation from diving. However, it could be said in overall terms that the inspection ROV is working in conjunction with, and in support of divers, by reducing the divers' workload and in-water time duration.

This is of course proposed from a technical aspect and not from the divers' paypacket point of view. It must however be said that ROV inspection, due to its increased efficiency, may well have increased the divers' workload by identifying areas requiring closer, more dextrous inspection.

Apart from this, very little has been achieved on a large scale in physical or tooling support of divers. It is perhaps an anomaly of the industry that work ROV and ADS/mini-sub projects have achieved considerable success as separate entities in support of subsea engineering tasks, and very little in support of divers directly.

There are a few excellent exceptions to this statement both past and present. Some immediate examples of Submersible technology in use with divers are the diver lockout submersibles, mobile diving unit (bells with 1 atm control) and the proposed DAVID ROV system.

FUTURE POSSIBILITIES AND DEVELOPMENTS

The present major advances in ADS/ROV technology are in advanced tooling programmes, which by definition has led to the development of larger, more capable vehicles to carry the tooling. The tooling advances are largely aimed at IMR structural operations. This advanced tooling capability could be made to work in conjunction with a diving operation. To draw a comparison, the capabilities of the proposed DAVID ROV system, designed primarily as an aid to diving operations, bear a surprising resemblance to the proposed capabilities of the new advanced ROV/ADS systems which are designed primarily as vehicle orientated systems.

The efficiency of ROV/ADS systems in visual/video and cp

structural surveys is I believe accepted by everyone in the industry. It is also a well documented fact that other work tasks including jacket cleaning using water jetting and grit blasting, more detailed cleaning and inspection using ultrasonics and MPI, etc. can be achieved with varying degrees of success. The important aspect to consider is the ongoing development that will make these tasks a definite reality and not a hit or miss operation.

Repair, maintenance and construction activities have seen advances in technology that allows tasks to be performed, such as the *in situ* replacement of anodes, that until recently have been considered the exclusive province of the diver.

The major advances made by ROV/ADS systems in recent years can be attributed to many aspects of technology; however, one aspect that has changed considerably, but is rarely documented, is the change in designer/operator philosophy toward the vehicles. Briefly, the change centres around the vehicle industry maturing away from a diver alternative system towards the recognition that the ROV/ADS system is a practical tool, a tool that can be changed and adapted to suit requirements, given the correct degree of operational and engineering input.

A COMBINED VEHICLE/DIVER SYSTEM?

A combined vehicle/diver system! Does this make sense financially or practically? If one considers that vehicle systems have not replaced divers, and are unlikely to do so in the near future, and that the vehicle system is a tool that can be adapted to perform specific tasks dictated by the degree of engineering input, then the answer is yes.

An example of an adaptable vehicle system is the modular concept Dragonfly ROV. This system has a base vehicle to which the operator attaches specifically designed modules to perform predetermined tasks. In its simpler base vehicle form it can conduct structural inspections with a level of efficiency equal to any present inspection ROV unit. By attaching a module it becomes a work ROV quite capable of supporting divers during a diving operation.

Recent design studies conducted for the Dragonfly vehicle would indicate that it is now within the capabilities of such a vehicle to supply to a diver the original eyeball surveillance, along with HP water jetting, hydraulic tooling, extra lighting and NDT packages including ultrasonic flaw detection and MPI; all in one vehicle with a single umbilical to the surface, removing the requirement for the diver to prepare the site or perform repeat bell runs for different operations and tools.

Given the required degree of engineering it is conceivable that the modular approach could also be utilized to provide a diver assistance unit for construction and repair tasks, including the carrying of heavy tooling and equipment.

CONCLUSION

In conclusion it is evident that the present ROV/ADS state of the art could be applied to the support of the diver to a greater extent than it is presently utilized and that the cost effectiveness of a combined activity could be demonstrated by increased efficiency.

The major problem is the persistent philosophy of treating the two activities as entirely separate, whereas in reality the ROV/ADS technology should be engineered to suit a diver-supportive role as a primary activity.

11

DAVID: A Remotely-controlled Multipurpose Vehicle Designed for Diver Assistance

E. Saunders, Shell UK Exploration and Production

SUMMARY

The tasks involved in the work of underwater maintenance, inspection and continued oil and gas field development require special equipment. This demand kindles initiative providing the incentive for the development of innovative labour-, time-, and hence, cost-saving devices. The system described in this paper is designed to provide a remotely operated underwater vehicle, primarily to assist the diver on location by providing all the tools and facilities, secondly to actually transport other tools to the worksite and thirdly as a vehicle capable of performing certain tasks in its own right as an ROV.

The vehicle system prototype, designed and manufactured by ZF-Herion Systemtechnik with the support of the Commission of the European Communities, has successfully completed a series of sea trials with Det Norske Veritas in Bergen, Norway. It is presently due for mobilization offshore onto Shell Expro's MSV Stadive to undergo trials, having been fully integrated into a specific work procedure.

INTRODUCTION

The growth of the offshore industry in recent times has led to corresponding demands on the diving industry and as a result diving techniques have been developed to a degree where the task of putting a working diver at a water depth of 200 metres has become routine.

The costs involved in operating a Diving Support Vessel (DSV) equipped with the necessary saturation diving spread are however high and this factor, together with the ever present risk of accident, has created a requirement for submersible machines which can perform the various work programmes, operated by remote control from the surface.

The commercial development of such machines has been continuous since about the mid-1970s and although a number of remote controlled and manned submersibles are already available and doing useful work, for many tasks there is as yet no real substitute for the diver working on site, particularly NDT and construction work.

This situation was recognized by ZF-Herion Systemtechnik and the company set out to design and manufacture a new concept in multipurpose underwater vehicles which could be employed either as a diver assistance vehicle or as a conventional ROV.

COSTS ASSOCIATED WITH DIVING OPERATIONS

A typical example of the equipment required to maintain a diver working in deep water would begin with a specially constructed Diving Support Vessel (DSV) of at least 2000 tons with accommodation for about 80 personnel. The vessel would carry a saturation diving chamber complex, a diving bell with appropriate handling arrangement and the various diving life support systems such as gas and hot water supplies. During diving operations it is essential that the DSV should maintain accurate position and heading above the work site and the vessel would therefore be equipped with a sophisticated dynamic positioning system.

The costs to charter such a vessel at present in the North Sea amount to approximately £20 000–25 000 each day that the vessel is on contract. Expressed another way, to transport and maintain a man at a subsea site for the purpose of carrying out

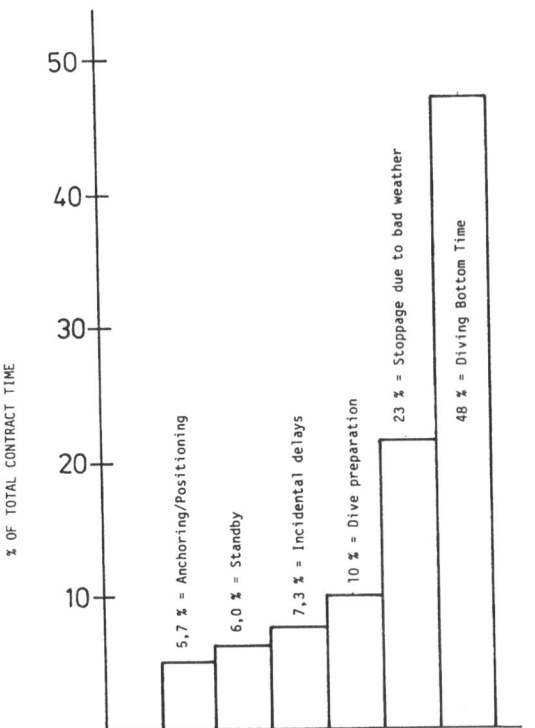

Fig. 11.1 Example of DSV operation. Total contract time = 500 hours

manual work tasks may cost may cost the oil company on average, around £1000 per hour.

Depending upon the particular contract, the diving bottom time (i.e. the time when the diver is available at the subsea work site to perform useful work), can be between 30% and 70% of the total port to port contract time. Figure 1 shows the breakdown of a typical contract in terms of time required for the various parts of the operation.

From this analysis, two points are immediately obvious. Firstly, because of the high rates for DSV operations, any reductions in the times required to complete contracts automatically result in significant savings with regard to costs involved. Secondly, the breakdown of the total contract time clearly shows that the area where effective savings are most likely to be realized is the diver bottom time.

CONDITIONS EXPERIENCED IN OFFSHORE DIVING

The environment in which the diver has to work is difficult and dangerous. The water temperature is generally between 4 and 7°C and natural illumination is either poor or non-existent. Tidal flow can be strong enough to prohibit diving completely and visibility is often drastically reduced by the particle content in the water. North Sea divers can be often working at depth of 150 metres, where they experience a pressure of sixteen times atmospheric pressure.

The 30 kg or more of life support equipment which the diver carries or wears is cumbersome and makes movement difficult and strenuous. The umbilical which tethers the diver to the bell usually consists of a hot water hose; two gas hoses and two communication cables, making a bundle of about 50 mm diameter. Problems can occur as a result of an uncontrolled ascent through the water and to prevent this the diver usually has a slightly negative buoyancy. Fins are not always worn when working in deep water at the sea bed or on structures, the divers sometimes preferring to use other boots for comfort and manoeuvrability.

In order to perform useful work tasks, additional tools and equipment must be made available to the diver and this may be done for example by lowering a basket from the support vessel above or rigging a shot line.

Apart from the problem of locating the basket, the diver then has the task of transporting tools and equipment from the basket to the worksite. The latter method involves rigging down lines to a suitable point of attachment. This preparatory work is essentially non-productive but because of the diver's limitations must be carried out, expounding energy, consuming time and exposing himself to the risk or delays of umbilical entanglement.

THE DAVID SPECIFICATION

Using reputable diving companies as sources of information, engineers of ZF-Herion Systemtechnik GmbH made a comprehensive study of the problems associated with diving operations. The findings indicated that there was a requirement for a vehicle which could be operated by the diver locally and which could provide tools and facilities as and when required by the diver, conceptual applications then formed a basis for the design specification.

The specification for the vehicle was greatly influenced by the demand for diving operations associated with the maintenance of jacket type structures where work is carried out by saturation divers working from a diving bell.

It was seen to be of particular importance that the vehicle should be already at the subsea worksite when the diver emerged from the diving bell, and a remote control facility was therefore considered to be essential.

Facilities which were seen as first requirements were:

- A submersible having a full ROV (Remotely Operated Vehicle) capability.
- A claw, adjustable in diameter, for clamping the vehicle to tubular structures.
- A moveable platform to provide a safe stable support for the diver.
- A source of power for hydraulic power tools.
- A range of underwater power and hand tools.
- A power winch, attached to the vehicle and with controls available for use by the diver.
- Adequate lighting.
- CCTV equipment.
- Cleaning equipment.

SAVINGS IN DIVING TIME USING THE DAVID

With the basic specification complete, the next step was to study actual work programs in order to assess the effectiveness of the proposed DAVID system. Independent diving specialists were contracted to define and analyse various subsea tasks and to compare task completion times using conventional methods with those achievable using the DAVID.

Seven examples, all considered to be routine tasks, were chosen and results were as follows:

Task	Diving time saved
MPI structural weld (saturation)	56%
MPI structural weld (air)	47%
MPI structural weld (seabed)	34%
Anode installation	51%
Installing riser clamp	58%
Pipeline repair	38%
Riser repair	37%

```
TASK:------- ANODE INSTALLATION ON INSIDE BRACE
WATER DEPTH: 75 METERS (250 FEET)
ALTITUDE:--- MIDWATER
DIVING:----- SATURATION, ONE DIVER WORKING FROM BELL
METHOD:----- LOWERING ANODE FROM SURFACE, TRANSFERING TO DIVERS,
             LIFTING GEAR, BOLTING IN POSITION
```

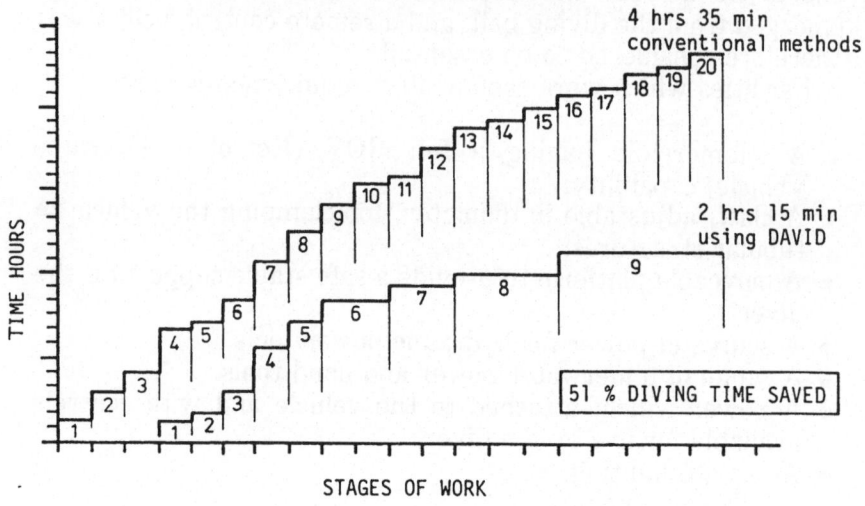

STAGES OF WORK

Conventional Method:

1. Locate and confirm position;
2. Establish downline and lights;
3. Send down rigging equipment;
4. Set up rigging;
5. Send down anode;
6. Transfer anode to rigging;
7. Position anode;
8. Install clamps and bolts;
9. Send down wrench;
10. Tighten bolts;
11. Wrench back to surface;
12. Remove rigging from anode;
13. Rigging back to surface;
14. Send down welding gear;
15. Weld cables to anodes;
16. Welding gear back to surface;
17. Send down video camera;
18. Video finished job;
19. Video camera back to surface;
20. Downline and lights to surface.

TOTAL TIME = 4h 35 min.

Using DAVID

1. Set up rigging;
2. Send down anode;
3. Transfer anode to DAVID winch;
4. Position anode;
5. Install clamps and bolts;
6. Tighten bolts;
7. Rigging back to DAVID;
8. Weld cables;
9. Video finished job.

TOTAL TIME = 2 h 15 min.

Fig. 11.2 Example of an offshore operation with a diver

Figure 2 shows a detailed explanation using the anode installation as an example. A general assessment based on these results alone must obviously be treated with care as contracts and conditions can vary considerably. Items not taken into account are the lost time due to bad weather and the aspect of diver safety. If tasks are completed in shorter times then lost time due to bad weather may also be reduced. These are bonus points which can only be really evaluated in practice, but can only increase savings.

Taking an average of 45% saving in diving time and relating this to the contract example previously described, it can be seen that using the DAVID may reduce the contract time from 500 to 392 hours for such a programme. This results in a cost reduction of around £100 000 on the contract and clearly provides commercial justification for the project.

It is appreciated that such savings are only realistic in terms of certain inspection or construction tasks. The saving on general visual inspection and cleaning will actually be less, basically saving diver time in going between basket and worksite which may account to less than 10% of an 8-hour shift.

On these marginal saving situations it was realized that it would only be a cost effective solution if the vehicle was developed as an ROV in its own right, capable of carrying out some cleaning and inspection tasks. This then became the design objective on the later developed model.

THE DAVID

The prototype system was manufactured according to the specification outlined earlier, and starting in the Summer of 1983 went through first commissioning in Lake Constance in Germany, through pilot and diver familiarization trials at NUTEC in Bergen, Norway, and latterly for operation in the North Sea on trial with Shell Expro proposed late 1984.

The design engineers responsible for the development of the system were actively involved in all stages of the trials so that both the concept and the design of the final product were subject to continual updating based upon operational experience.

An outline drawing of the final design is shown in Fig. 3 and this illustrates how the specification was realized.

This modified vehicle is due for water trials in early 1985.

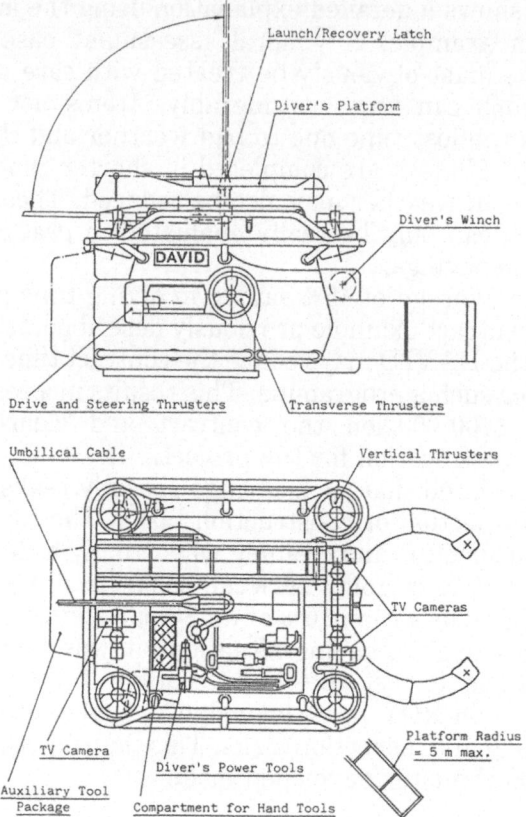

Fig. 11.3 The DAVID diver assistance vehicle

Control and Navigation System

A fundamental requirement for the concept is that the surface operator must be able to pilot the submersible to the subsea worksite and there to dock onto a structure at any chosen position. Further to this it is designed such that the diver on site may then make minor adjustment to position using the remote control box facility. The control system was designed to provide the degree of stability required for this phase of the operation.

An onboard computer receives actual value input signals from sensors indicating depth, roll, pitch and heading, and set value signals from the surface control station. The sensor inputs are used to calculate the actual vehicle orientation and this is then compared with the required orientation derived from the control station inputs. The difference is evaluated and outputs to the thruster speed regulators effect the required correction. From the surface operator's point of view, the vehicle remains stable in attitude, depth and heading, and commands from the steering joystick are superimposed. Commands to travel vertically upwards or downwards are derived from push-buttons and movement in this direction is made at a constant speed irrespective of load on the vehicle.

Claw Arrangement

The claw is designed to attach to the tubular members of offshore structures and can be aligned either by rotating or tilting as the vehicle approaches the chosen docking position. The claw pressure pads grip the tube at points which are approximately equally spaced and the mechanism functions using a single hydraulic cylinder. In order to achieve a compact design, diameter setting is provided in two ranges, the range adjustment being made by the surface operator as required.

Platform Design

Because the diver is virtually weightless in water it is difficult for him to work with equipment which is not itself weightless. Without some kind of support there is no way that he can exert pressure when using tools.

The platform is a ladder type arrangement and is dimensioned so that the diver can obtain a firm anchorage using only the legs. Both hands are then free to work with equipment.

Tools and Equipment

A set of power tools and a range of small hand tools are carried on the vehicle as standard equipment. The hydraulic power supply for these tools is separated from the main system. Additional heavy equipment units can be mounted on the rear of the vehicle and each of these units is trimmed to be neutrally buoyant. Connection is made into the main hydraulic supply before the vehicle is launched.

Power Winch

In order that the diver can use the winch, the submersible must first be clamped onto the structure by means of the claw. The line of action is then from the winch through the claw and into the structure itself. The obvious mounting point for the winch therefore is on the claw frame. The winch/claw arrangement has been load tested for a safe working load of 1.5 tonnes.

Lighting and TV Equipment

The vehicle is equipped with five separate lighting circuits. The TV cameras are mounted on pan and tilt units and cameras can be removed by the diver for close-up observation. The pan and tilt units are controlled from the surface.

PROJECT STATUS AND CONCLUSIONS

The object of Shell Expro's trial is to assess the commercial viability of the system and validate the estimates for saving in diving times.

The contribution offered by DAVID to the offshore industry should be threefold:

1. Costs of maintenance and repair work are significantly reduced for the oil company; operators and service companies achieve greater productivity.
2. The diver receives a powerful and flexible tool which reduces physical exertion.
3. Conditions for safety at the worksite are improved in view of stability, light, communications and reduced exposure.

The vehicle provides a new approach in the ongoing development of ROVs and, more so, the associated workpackages.

To date development of cleaning and inspecting equipment has evolved as a result of a vehicle being utilized in an increased work capability — at this stage quite often a market exercise would be carried out to promote the new capability.

The DAVID vehicle offers a more analytical approach to the same development. Through a careful study of the use of the DAVID on the Trial workscope, and beyond this other working contracts, the use of various items of equipment may be studied regarding frequency of use and a critical assessment of their practicality, the end result being a tool custom designed from operational feedback, hence the development of a working ROV from the working tool backwards to vehicle specification for the most effective solution.

In its prototype form the vehicle presents difficulties in incorporating into a DSV work programme. As an unknown quantity its deployment has to be carefully planned into the work procedure, as opposed to an *ad-hoc* usage. To leave the tool available to a diving contractor at this stage may be unproductive; due to commercial pressures the company would stand a risk of inefficient performance as co-ordinated operation may be difficult on the early part of a learning curve.

Once the vehicle is actually seen to be working it should stimulate the operators' imagination and its versatility be exploited into a wide range of opportunities — this familiarization process and change in approach is an important stage in obtaining maximum utilization of the vehicle as an extension of the diver himself.

The situation was appreciated by Shell and ZF-Herion and for this reason they believed it was the best trial arrangement for fully evaluating the vehicle away from the restriction of commercial pressures.

12

Effects of Saturation Diving on Respiratory Systems

Enrico M. Camporesi, MD, John V. Salzano, PhD, Richard E. Moon, MD, FRCP and Bryant W. Stolp, BS, Duke University Medical Center, Durham, USA

INTRODUCTION

Physiological limitations encountered during saturation diving are usually multifactorial. Respiratory problems, which represent only one facet of the difficulties encountered, have received wide attention in the last few years, since various ventilatory limitations have been observed during work at depth.

The respiratory system can be affected by saturation diving as a result of alteration of one or more of the various components of the environment which include: hydrostatic pressure, gas density, oxygen pressure and partial pressure of the inert gas species. Effects of abnormal oxygen pressures, hypoxia or O_2 toxicity, can usually be controlled by manipulation of the O_2 content of the gas mixture. The effects of gas density and pressure *per se* may act singularly or in combination as agents which may modify the functions of the respiratory system. Neural and chemical control of the system may be depressed or potentiated, depending on the absolute pressure.

The debilitating effects of high pressure nervous syndrome (HPNS) interfere with the ability of man to perform useful work at great depths. It is possible to attenuate HPNS by adding nitrogen to the helium-oxygen breathing mixture (heliox) to counterbalance the effect of rapid changes in pressure and/or of pressure *per se*.[1] However, the addition of nitrogen to heliox necessarily increases the density of the breathing gas. The net effect, therefore, might be an amelioration of HPNS symptoms accompanied by a gas density dependent reduction in ventilatory ability, a reduction in gas exchange reserve and of work capability.

Increased gas density has definitive effects on the mechanics of the system[2] and may also affect performance of the respiratory muscles. Maximum expiratory flow decreases in a somewhat predictable fashion as a result of flow resistance changes caused by an increase in gas density, while fatigue of inspiratory muscles can influence ventilatory volumes. Efficiency of gas exchange may be improved by better distribution of the dense gas within the lung relative to distribution at normal density. On the other hand, poor intra-airway mixing or stratified inhomogeneity may adversely influence gas exchange. Delivery of O_2 may be compromised by an effect of pressure on hemoglobin affinity for O_2. Acidemia from retention of CO_2 and increased arterial lactic acidosis are other results of an increase in gas pressure and/or density. The interaction of all factors on work capacity is not predictable. Quantitative descriptions of these changes constitutes the major portion of this report.

METHODS

All data presented in this paper were collected during the past five years during deep dives in a dry chamber, involving three subjects per dive.

Studies were conducted while breathing a variety of gas mixtures with densities ranging from 1.1 to 17.1 g/l (BTPS). The subjects breathed either heliox or a mixture of heliox and nitrogen (trimix) at pressures of 1509 feet sea water (fsw) or at 2132 fsw.

All gases contained O_2 at a pressure of 0.5 ATA. Table 1 contains a more complete description of the depths and gas compositions used in this series which have been named the Atlantis Dives. The subjects varied in backgrounds ranging from professional divers (SP and LW) to a medical student (WB). Lung capacity and other gas exchange capabilities of the subjects are shown in Table 2.

The subjects spent several weeks before each dive in a training and conditioning program. All subjects improved their work

TABLE 1
Control and experimental conditions during Atlantis I, II and III dives

Inspired Gas	Pressure (ATA)	Density (g/l) BTPS	Depth (msw)	PIO_2 (ATA)	PIN_2 (ATA)	Atlantis Dive No.
Air	1	1.1	0	0.21	0.79	I,II,III
O_2/N_2	1	1.2	0	0.50	0.50	I
Heliox	46.7	7.9	460	0.48	0.00	I
Trimix-5	46.7	10.1	460	0.50	2.34	I,II
Trimix-10	46.7	12.3	460	0.50	4.67	II
Trimix-10	65.6	17.1	650	0.50	6.56	III

TABLE 2
Subjects in Atlantis I, II and III dives

Subject	Age (years)	Height (cm)	Weight (kg)	VC (l BTPS)	1 ATA MVV ($l\ min^{-1}$ BTPS)	1 ATA VO_2 max ($l\ min^{-1}$ STPD)	Atlantis dive no.
WB	25	183	71	6.23	253	3.10	I,II
SP	24	185	87	7.36	244	4.50	II,III
DS	40	175	79	4.87	195	3.60	I,II
LW	27	173	77	5.20	230	2.95	I,III
EK	24	183	74	6.13	214	3.20	III

capacity during this conditioning period, as evidenced by progressive reductions in heart rates during submaximum work on a cycle ergometer. Data for comparison of work performance at surface and at depth do not include those from this conditioning period. In addition to physical conditioning during the predive period the subjects learned to perform a variety of experimental procedures such that they became both subject and experimenter during the studies at pressure.

Three subjects participated in each dive; all but one was a subject in two dives; none participated in every dive. Two members of each dive team were trained to insert arterial cannulae for blood sampling. This procedure was performed successfully on each diver in all three dives.

An open circuit system with a low-resistance breathing valve was used to deliver gas to the subject and for collection of exhaled gas. Gas for inspiration was from a supply separated from the chamber gas in order to ensure uniform gas composition during any series of measurements. Inspiratory flow rates and volumes were measured with a pneumotachometer in the inspiratory line. Expired gas volumes were collected for one minute intervals, in large plastic bags. The volume in the bags was measured with a dry gasometer inside the chamber at the same depth as it was collected. Samples of inspired and expired gases were collected through special sample-lines outside the chamber and measured either chromatographically (CO_2) or with a fuel-cell O_2 analyzer. Arterial blood samples were collected anaerobically during the sixth minute of a given exercise period, placed in ice and later measured with an appropriate blood–gas electrode system which was located inside the chamber. Exercise was performed on a cycle ergometer with pedal rates of 60 revolutions per minute and at various loads.

Arterial lactate and pyruvate concentrations were measured in samples collected at rest and one minute after each six-minute exercise level during surface control and depth, during Atlantis III only. The blood samples were rapidly mixed with perchloric acid to stop enzymatic reactions, decompressed in a controlled manner and the resultant supernatant was later analyzed at 1 ATA with standard techniques.

Hb dissociation curves of blood samples drawn from divers at pressure were measured directly in the hyperbaric chamber during the decompression stages of Atlantis IV. During this last dive we could not complete exercise studies since one of the new subjects exhibited prolonged excitation during the last day of compression to 65.6 ATA, leading to a reduction of the scientific program. A modified spectrophotometer was used to determine the oxygen dissociation curves of whole blood from three subjects.

Standard equations were used for $\dot{V}E$, $\dot{V}O_2$ and $\dot{V}CO_2$ calculations. Statistical analysis of the data was approached with a worst-case error analysis, since traditional tests could only be applied sparingly due to the limited number of replications.

RESULTS AND DISCUSSION

Data were analyzed in several ways. Firstly, in order to determine if the PO_2 of 0.5 ATA used at depth would be responsible for any differences, a comparison was made between results obtained at 1 ATA while the subjects breathed air and while they breathed 50% O_2 in N_2. Additionally, some experiments at 1 ATA were carried out in the morning and these were compared with similar procedures performed in the afternoon in order to mimic the schedule for measurements at depth. No consistent, systematic variations in results could be assigned to differences in PO_2 or time of day at 1 ATA. Therefore, all measurements on a given individual at 1 ATA were pooled to form a surface or 1 ATA value for that diver at rest and at several work rates. There was no systematic difference among divers; therefore, the results from each diver were pooled to form a 1 ATA value. Data collected during exposure to increased pressures were not affected in any of the divers by changing the gas density from 7.9 g/l with heliox at 47 ATA, to a density of 17.1 g/l with trimix-10 at 66 ATA (10% N_2 in heliox). Consequently all pressure values were pooled at each of three or four work rates and the mean of these is referred to as 'depth' value, irrespectively if collected at 47 or 66 ATA, or breathing a gas with density of 7.9; 10.1; 12.3 or 17.1 g/l. These results are presented in greater detail elsewhere.[3]

Arterial PO_2 (PaO_2) measured at depth at rest and during all exercises was never lower than 250 Torr, assuring saturation of arterial blood in all conditions studied.

Work rates of 360 kpm/min (60 watts) were successfully completed for six minutes by every subject. Most were able to work at rates of 720 and 900 kpm/min. One subject (SP) was able to work at 1440 kpm/min for 5 minutes while breathing trimix-10 (17.1 g/l) at 66 ATA. This is the only demonstration of man's capability for such severe work at these extremes of depth and gas density. A quantitative description of the physiological responses to work at 1 ATA and at depth is presented in the following figures. For the sake of comparison, a line of identity (broken line), is shown to indicate no difference between 1 ATA and depth. Actual measurements at surface and during comparable work rates at depth are shown by closed circles. The pattern of deviation from identity is shown by the solid line which connects the points. Deviations from the identity line are thus highlighted.

Figure 1 summarizes the steady-state results for oxygen consumption (VO_2) and CO_2 production (VCO_2) during rest and at increasing work levels at 1 ATA and at depth. Metabolic rate increased about ten-fold from rest to the maximal work level performed (1440 kpm/min), with only minor deviations from the identity line.

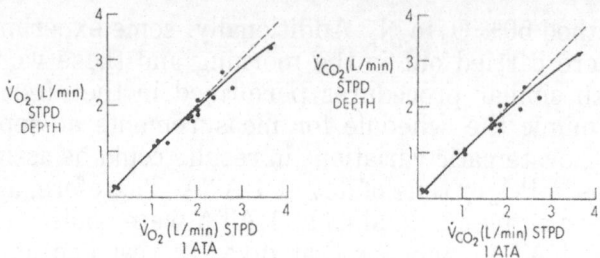

Fig. 12.1   Pooled data for $\dot{V}O_2$ and $\dot{V}CO_2$ obtained at 'depth' (see text) plotted as a function of corresponding 1 ATA value. Each iso-work point represents the average of several replications

At rest and during work rates up to 720 kpm/min the pulmonary ventilation at depth was equal to or greater than measurements made at the surface. At the highest work rates attempted by each subject the pulmonary ventilation at depth was consistently less than corresponding 1 ATA values (Fig. 2). During 1 ATA control, ventilation values at work rates greater than 720 kpm/min increased out of proportion to an increase in work rate, suggesting the onset of an anaerobic threshold. At depth the phase of this accelerated ventilation was not as clearly evident.

Individual values of alveolar ventilation ($\dot{V}A$) were calculated from $PaCO_2$ and $PECO_2$ values using the Enghoff modification of the Bohr equation. Figure 3 shows the comparison of alveolar ventilation values for iso-work points at depth and at the surface.

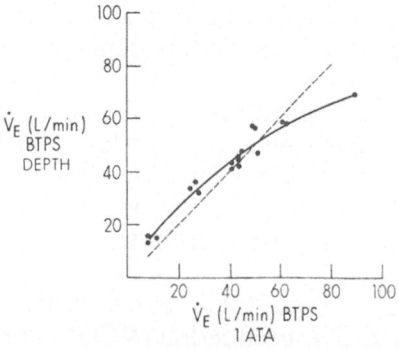

Fig. 12.2   Pulmonary ventilation ($\dot{V}E$) during rest and work (6th min) at depth and at 1 ATA. Deviations from the identity line indicate hyperventilation at depth for moderate work rates, and hypoventilation at high work rates

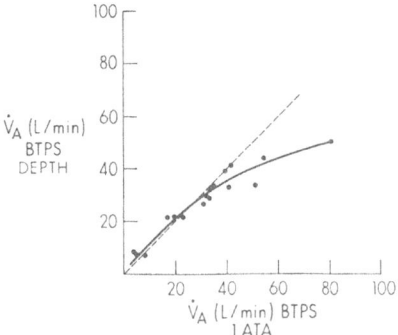

Fig. 12.3   Alveolar ventilation ($\dot{V}$A) at depth and at 1 ATA

At rest and during mild work rates $\dot{V}$A at depth was indistinguishable from 1 ATA control. At higher work rates, as suggested by the marked hypercapnia observed, alveolar ventilation at depth was greatly reduced compared to similar work rates at 1 ATA for each subject. Figure 4 summarizes our findings in a more schematic manner.

Arterial lactic acid concentration at rest and during the first minute of recovery from each level of work was greater at depth than at the surface. At 1 ATA, the arterial lactic acid concentration at 360 and 720 kpm/min was not significantly different from rest whereas at pressure the lactate concentration at

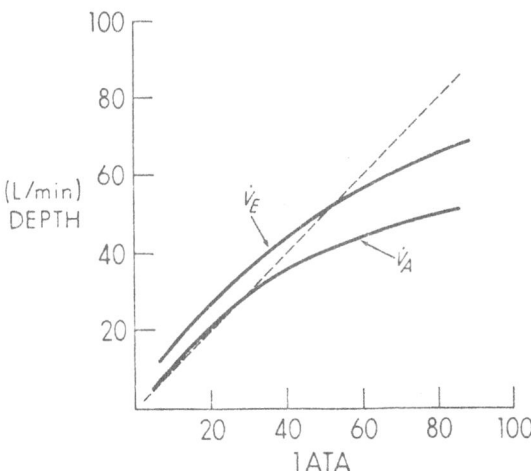

Fig. 12.4   Best-fit line of regression for all subjects for $\dot{V}$E and $\dot{V}$A, from rest to maximal work rate, comparing depth values to 1 ATA control. The axis values refer to both $\dot{V}$E and $\dot{V}$A

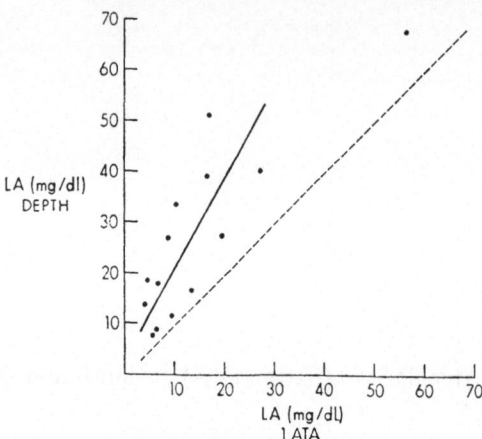

Fig. 12.5   Arterial lactate concentration at rest and after work at depth
and at 1 ATA

720 kpm/min was significantly higher than either at rest or at
360 kpm/min. Hence, the onset of lactic acidosis at pressure
occurred at a lower work rate than at the surface. Figure 5
illustrates the much higher lactate levels measured after work
at depth, compared to 1 ATA controls. These data were only
obtained during the Atlantis III studies.

As indicated above, arterial oxygenation was higher at depth
(PIO_2 = 0.5 ATA O_2) during Atlantis III than during 1 ATA
control (air). Despite this, much higher lactate levels were
observed at depth. This novel observation may be indicative of
limitations of aerobic metabolism at depth, or may result from
an insufficient metabolism of this compound, relative to its
evolution in blood. Our data do not allow a reasoned selection of
the causative agent of the higher lactate levels observed at
depth. However, the consequences of these higher values is
reflected in the larger degree of metabolic acidosis at depth,
during moderate to heavy work rates. This alteration of exercise homeostasis is well summarized in Fig. 6, which compares
$PaCO_2$ and arterial pH values obtained at the surface to those
observed at depth.

The arterial PCO_2 ($PaCO_2$) at 1 ATA was maintained during
low work rates at the same level as at rest. A further increase in
work rate was accompanied by a decrease in $PaCO_2$ (above
anaerobic threshold 'blow-off' phase). At depth, a markedly different pattern of response was observed: with increasing work

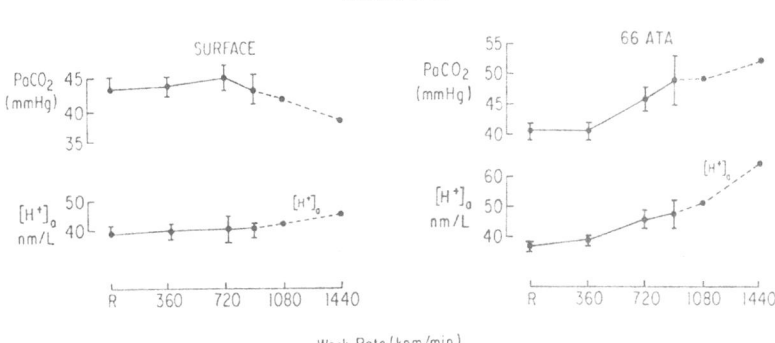

Fig. 12.6 Arterial PCO_2 and pH (in nM/l) at 1 ATA and at depth for all subjects as a function of increasing work rate. Broken line joins points with less than three replications

rate arterial hypercapnia was observed where normo- or hypocapnia occurred at the surface.

This last figure shows how the increase in pulmonary ventilation at 1 ATA maintains a constant pH up to very elevated work rates. At depth pH cannot be maintained equal to resting values: arterial $[H^+]$ starts climbing above resting values already at moderate work rates, and arterial hypercapnia is evident during heavy work. Our subjects at depth demonstrated both a metabolic and a respiratory acidosis, in contrast to the normal response to heavy exercise exhibited at 1 ATA, where respiratory alkalosis compensates almost completely the metabolic acidosis.

Dyspnea was never reported at 1 ATA, but unpleasant respiratory sensations at depth were reported by all subjects. In general, dyspnea was experienced as a sensation of insufficient inspiratory power, and it increased in severity with the increasing intensity of work. Some subjects reported expiration was also impaired. Dyspnea was reported in 24 of 54 exercises, but only in five cases was it considered severe enough to induce premature work termination at depth.

Analysis of the MVV15 data obtained before exercise at 1 ATA, at the maximal pressure of exposure, and during decompression, during each dive demonstrated a decay of MVV15 (inverse power function) with increasing gas density. In several instances larger values of MVV were obtained at pressure by some subjects utilizing a lower breathing rate than at the surface. Additionally, some subjects were able to sustain for several minutes exercise ventilations nearing, and sometimes exceeding MVV15 values. The relative inadequacy of VE to maintain normocapnia during heavy work, however, due to the profound reduction observed in VA at pressure, renders predictions of exercise performance based on MVV data limited in value. These observations have led us to believe sustained ventilation may be a better predictor of ventilatory capacity at depth.

The O_2 pressure at 50% saturation (P-50) for hemoglobin was obtained during decompression of the subjects of Atlantis IV from 55 ATA to the surface. The modified Hemox analyzer was used to obtain hemoglobin dissociation curves under the following four conditions:

(A) *Pre-dive control* blood samples were obtained at 1 ATA and analyzed at 1 ATA with the cuvette isolated in the chamber.
(B) *Decompressed* blood samples were drawn from venipunctures during compression depths between 530 and 650 msw (54 and 66 ATA). Due to space and scheduling considerations, the isolated cuvette and gas exchange apparatus could not be placed in the chamber during this stage of the dive. Therefore, the blood samples were decompressed to the surface (20 minutes) using fluorocarbon FC-80 as a gas sink. For these determinations, the cuvette was placed outside the chamber at normal barometric pressure.
(C) *Depth* measurements were obtained during a 22 day period of decompression from 530 to 120 msw. Fresh blood samples

TABLE 3
P-50 values (Torr, ± 1SD) for three divers in Atlantis IV. Numbers in parenthesis indicate replications

	Pre-Dive A	Dive B	Dive C	Post-Dive D
P-50 (Torr)	26.8 ± 1.5 (9)	26.9 ± 1.9 (25)	24.4± 1.0*(16)	27.6 ± 1.2 (13)

A – Samples collected and analyzed at 1 ATA.
B – Samples collected at depth and analyzed at 1 ATA.
C – Samples collected and analyzed at depth.
D – Samples collected and analyzed at 1 ATA.
*$P < 0.05$: significantly different from pre-dive control.

were drawn from finger sticks and analyzed with the isolated cuvette *at pressure*.

(D) *Post-dive* samples were drawn for five days after the dive
and analyzed as described for pre-dive controls.

The results for the four conditions are reported in Table 3,
and individual data have been published in more detail.[4]

These results for the 63 complete dissociation curves acceptable for analysis indicate that P-50 is significantly lower *at pressure* in the range of 120 to 530 msw. The physiological impact of
such a leftward shift at pressure is, however, debatable. Given
that arterial PO_2 values in our subjects were high, a leftward
shift in the curve would result in lower oxygen availability for
any tissue blood flow and venous PO_2. However, several com-
pensatory mechanisms seem to be available in the working muscle to offset this small reduction in P-50.

CONCLUSIONS

The data from the Atlantis dives revealed new limits to human
performance: dyspnea, early arterial acidosis, blunting of the
ventilatory homeostatic mechanism and rapid deterioration of
ventilatory efficiency as evidenced by much larger physiological
dead space at depth. It is unclear which limitation may restrict
maximal physical performance at moderate exercise levels.

In summary, various physiological alterations appear to be
operative at depth, and contribute to various kinds of work
limitation. Future research must clarify the role of muscular
insufficiency of the respiratory apparatus while breathing dense
gas. Recent observations suggest that signs of diaphragmatic
fatigue are more rapidly demonstrated during arterial hypercapnia compared to normocapnia, and that elevated CO_2 levels
adversely affect resolution of fatigue.[5] A growing body of evidence also suggests that various training and pharmacological
regimens may increase respiratory muscle endurance,[6] and
studies are needed to decide if training of respiratory muscle
may increase exercise tolerance in divers.

The large dead spaces operative at depth appear more unsur-
mountable. Our data show a strong correlation at depth between dead space and the tidal volume used during exercise.
Optimization of breathing pattern during work at depth may
permit selection of frequencies of breathing and tidal volumes

which may maximize ventilatory volumes and minimize dead space ventilation.

Our finding of a significant, though small, reduction of P-50 at depth needs to be confirmed and studied with a protocol extending the range · of the observation between 50 and 100 ATA. More pressingly, the physiological relevance of this alteration must be established *in vivo*, a task which may require exposure to pressure of an animal model.

We believe today that actual measurements of maximum work tolerance at increased pressures may still reflect an optimistic view of work capability. Such measurements may ignore important derangements, such as metabolic acidosis, which may exceed the capacity of the body to provide acute, and especially long-term, compensation. As previously mentioned, our data were collected in near optimal conditions and in a dry simulation chamber. Saturation exposures to great depths will require long durations of compression and stages, and field-type conditions which may only further reduce the available reserves for safe human performance.

REFERENCES

1. P. B. Bennett and M. McLeod, 'Probing the limits of human deep diving'. *Phil. Trans. R. Soc. Lond.* **B 304**, 105–117 (1984).
2. C. J. Lambertsen, R. Gelfand, R. E. Peterson, R. Strauss, W. B. Wright, J. G. Dickson, C. D. Puglia and R. W. Hamilton, Jr., 'Human tolerance to He, Ne and N_2 at respiratory gas densities equivalent to He-O_2 breathing at depths to 1200, 2000, 3000, 4000 and 5000 feet of sea water'. *Aviat. Space Environ. Med.* **4**, 843–855 (1977).
3. J. V. Salzano, E. M. Camporesi, B. W. Stolp and R. W. Moon. 'Physiological Responses to Exercise at 47 and 66 ATA'. Accepted for Publication in *J. Appl. Physiol.* **57** (1984).
4. B. W. Stolp, R. E. Moon, J. V. Salzano and E. M. Camporesi. 'P-50 in divers decompressing from 650 msw'. Underwater Physiology VIII, Proceedings of the 8th Symposium on Underwater Physiology (Eds.) A. J. Bachrach and M. M. Matsen, Underwater Medical Society, Bethesda, Maryland: 315–326 (1984).
5. E. J. M. Campbell, 'Discussion: acidosis and tension development. Human Muscle Fatigue: Physiological Mechanisms'. Ciba Foundation Symposium 82, London: 82–88, (1981).
6. D. Murciano, M. Aubier, Y. Lecocguic and R. Pariente. 'Effects of theophylline on diaphragmatic strength and fatigue in patients with chronic obstructive pulmonary disease'. *New Eng. J. Med.* **311**(6), 349–353 (1984).

13

Long-term Effects of Professional Diving

R. I. Mc'Callum, University of Newcastle
upon Tyne.

SUMMARY

The study of possible long-term effects of diving is limited by
certain constraints including definitions (subject, diving exposure, etc.) and other variables. The range and intensity of
diving activity in terms of time, frequency, depth, gas mixtures,
etc. is large and the same individual may be exposed to a wide
variety of conditions over a working lifetime. Divers themselves
are highly self selected and the nature of their work makes
epidemiological follow-up for scientific purposes difficult.
Nevertheless bone necrosis has been successfully investigated
in a group of professional divers and the study of even small
numbers of divers with neurological defects associated with diving is likely to produce important information. The Decompression Sickness Central Registry in the University of Newcastle
upon Tyne is in a unique position to contribute to the long-term
study of professional divers but there are difficult problems arising from current economic pressures and the need to maintain
good records over a sufficiently long period of time.

INTRODUCTION

Concern about possible long-term effects of professional diving
has grown over the last few years and intensified as deep diving

techniques have developed. Early diving legislation in the United Kingdom[1] required medical examination and a chest radiograph but unfortunately there was no mechanism for evaluating the results of these examinations so that no published useful information has emerged from them. With the beginning of the rapid development of commercial diving associated with the discovery and exploitation of gas and oil in the 1960s the main concern was with accidental deaths and the acute risks of diving, particularly in North Sea conditions. In 1975 The British Medical Association's Scottish Council published a widely read report[2] which attempted to define the hazards of diving and drew attention to the then high mortality in North Sea divers. The adverse conditions which prevailed at that time have as a result largely been controlled. Nevertheless it was seen that modern commercial deep diving had evolved a long way from the type of diving which had been useful and fairly static over the long period since August Siebe perfected the diving helmet in 1837. The possibility of unknown long-term effects as a direct result of new diving techniques, especially saturation diving on mixed gas, began to be considered. In 1978, for example, the Mines Safety and Health Commission of the Commission of the European Communities held a Workshop in Luxembourg on 'Long-term Health Hazards of Diving' which included papers on the central nervous system,[3] bone, the internal ear, chromosomes, carbon dioxide toxicity and metabolic changes.

Some of the concerns of 1978 are still unresolved and a cause of worry but others have receded in importance since then. But in few or perhaps none of these hazards is it possible to be satisfied that adequate solutions have been arrived at.

In evaluating factors which might in the long term be likely to cause permanent damage and in attempting to assess what if any changes might result from them, certain difficulties present themselves which limit the acquisition of good data on which to base conclusions. For example, tissues such as bone and nerve react in a relatively limited number of ways to a wide variety of harmful agents, so that unless highly specific changes can be identified and linked with particular diving exposures epidemiological techniques must be used. This means comparing groups of divers with similar work experience, with control groups on non-divers matched for age, sex, obesity and so on. The nature of diving activity makes it difficult to set up and carry out a satisfactory survey and presents too many variables in diving experience. There are, for example, disparate patterns of shallow air diving and deep mixed gas diving; different decompression schedules at different times in a career and marked differences in intensity of diving. As divers are highly self selected and therefore in some ways an 'abnormal' group it is difficult to know who best to compare them with. Where a long-term hazard is under consideration problems of definition occur, for example, how long is 'long term' and how long is a diving career and what sort of career should one choose? Furthermore the majority of divers appear to stop active diving or change their pattern of diving[4] at around 35 years of age. Furthermore what weight does one give to air diving as opposed to mixed gas diving when the former is thought by many people to be much more hazardous?

The problems to which attention has mostly been given so far are: bone necrosis, neurological damage and chromosome changes.

BONE NECROSIS IN DIVERS

Death of parts of long bones has been the subject of intense epidemiological and experimental work since the 1960s when it was realized that in compressed air workers symptomless bone

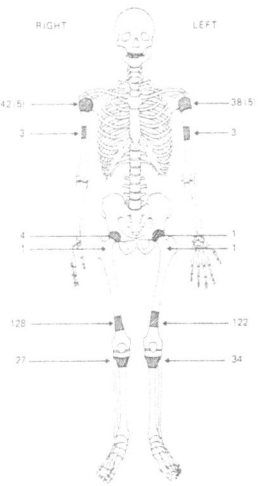

Fig. 13.1 Sites of bone necrosis in 247 divers showing number of lesions at each site, and in brackets the number of damaged joints. The great majority of lesions are symptomless and not disabling, and will remain so

TABLE 1
Bone radiographs of commmercial divers to July 1984 [a]

Radiograph		*No. of men*	
Normal		3358	
Head, neck, shaft lesions		216	(60)
Juxta articular[b] lesions		77	(37)
Irrelevant changes		3213	
	Total	6961	

[a]Out of 6961 divers whose radiographs have been scrutinized at the Decompression Sickness Registry, 293 men have one or more areas of bone necrosis and 97 have suspected lesions. Many others, mainly trivia abnormalities unconnected with diving, are also found ('irrelevant changes').
[b]Disability in 12 men (Suspected lesions in brackets)

damage could be found in men currently at work in raised atmospheric pressure. It has since been demonstrated[5] that divers also have symptomless bone damage but to a lesser extent (overall about 4% compared with about 20% of compressed air workers). Important in professional divers engaged in well controlled diving is the rarity of the disabling forms of bone necrosis, particularly that occurring in the femoral head (Fig. 1), compared with compressed air workers. Nevertheless in men who have dived relatively deeper than others the proportion with bone damage may rise to 20 or 30%. To a great extent the problem of bone necrosis in divers has been well defined in relation to the technique and exposure used commonly in the North Sea over the last 10 years. It is numerically small and largely symptomless but not negligible (Table 1).

The importance of symptomless bone necrosis in divers is that its occurrence demonstrates that existing compression/decompression procedures cannot be regarded as satisfactory quite apart from the prevalence of acute decompression sickness. For this reason changes in diving techniques, for example, air saturation diving, need to be monitored especially carefully to determine how safe they are in terms of bone damage.

NEUROLOGICAL AND PSYCHOLOGICAL EFFECTS

Long-term damage to the central nervous system may be evident following acute decompression sickness (Type II) which has not fully recovered. Historically, diver's palsy has been well known for many years[6] and the site most commonly affected is the spinal cord.[7] Paraplegia is a common risk of relatively shallow diving on air, probably much more than in deep diving with mixed gas, and is often seen in sports divers.[8] Although treatment can be expected to give a good functional recovery there may be evidence of permanent damage subsequently[9] and extensive cord damage may be present with very little in the way of abnormal signs on neurological examination.[10]

The possibility has been raised that frequent diving *per se*, particularly saturation diving, without neurological decompression sickness or indeed any history of decompression sickness might give rise to repeated damage to the cerebrum or spinal cord, with little in the way of abnormality on routine neurological examination. The same question was raised in connection with the use of compressed air in the development of Budapest's underground railway system which began in the 1950s and in which 6000 men were exposed to 18–25 psig pressure[11] over a period of 15 years. The decompression sickness rate was 2% overall and of these 2.72% had 'central neurological forms', and 3.5% Menière's Syndrome. The neurological decompression sickness is described as having no systematic signs but multifocal deviations from normal, affecting the cerebrum, upper brain stem, medulla, pons and cerebellum as well as the spinal cord. Psychiatric changes are also described, in the form of a vegetative neurosis (neurosis with symptoms of autonomic dysfunction); late onset impotence, personality changes including loss of self-control, intolerance to alcohol and pathological drunkenness, sleep disorders, loss of mental ability, and lassitude, anorexia and headaches; nystagmus and abnormal skin reflexes are also described. It is emphasized that the sensitivity of methods of examination is important in detecting minor defects.

Ròzsahegyi states that damage to the central nervous system during decompression can be latent and without any decompression sickness. In 31 compressed-air workers without a history of decompression sickness there were nine pathological and four borderline electroencephalograms (EEGs) and the abnormal findings were linked to chronic psychological disturbance. Unfortunately this work has never been repeated elsewhere and the observations have not been confirmed by a controlled study. Although there are difficulties in accepting some of the statements, arising for example out of conceptual differences between Eastern European physicians and those in the UK in describing psychological states, it has not been possible to ignore Ròzsahegyi's work.

Concern about the possibility of similar long-term changes in divers has existed for some time and has been intensified by the occurrence of the High Pressure Nervous Syndrome (HPNS) in deep diving using helium/oxygen as the breathing gas.[12] Decreases in motor and intellectual function together with dizziness, nausea, vomiting, tremors are encountered during deep dives on helium and oxygen, sometimes with EEG changes, and are related to speed of compression. All these phenomena can be suppressed by addition of nitrogen and are reversible.

In 1978 Hallenbeck[3] commented on the absence of 'hard' data relating to the effects of diving on the central nervous system (CNS). He described three levels of injury: (a) CNS damage without a history of either decompression sickness or air embolism; (b) neurological decompression sickness followed by progress CNS degeneration; and (c) CNS damage related to continued diving in individuals with a history of decompression sickness. In the first the effect of multiple silent bubbles during decompressions is postulated but clinical symptoms would be expected with focal ischaemic damage, and in any case recovery from this would be expected with time. Recovery would also be expected ultimately in the second example, while the third implies further injuries to nerve tissues already damaged and perhaps rendered more susceptible because of defective blood supply. Hallenbeck concluded that at the time there were too few firm scientific observations to answer the questions being raised.

Since then, diving has continued to develop, saturation diving has been to greater depths and with it concern over possible deleterious effects has increased. For example, there are anecdotal accounts of behavioural changes in some saturation divers and of the development of intolerance to alcohol, reminiscent of Ròszahegyi's account of compressed-air workers. The changes alleged to occur in divers have come via their wives and an attempt is being made to investigate their claims.

In November 1983 the European Undersea Biomedical Society and the Norwegian Petroleum Directorate organized a Workshop on 'The Long-Term Neurological Consequences of Deep Diving'[13] in response to anxiety which had arisen as the economics of oil exploration and exploitation demand more from diving and divers and which might be exposing them to unknown hazards.

The meeting reviewed the neurological effects of deep diving, particularly in relation to the compression phase and the stay at pressure. It was recognized that two factors needed to be separated: the possible long-term effects of high gas pressure itself and secondly, the effects of decompression from high pressure although making a clear distinction between them might not be possible.

It was evident that a better understanding is still required of the effects on the central nervous system of high ambient gas pressure, confirming the persistent absence of hard data referred to in 1978. The high pressure nervous syndrome, and how it relates to electroencephalographic changes and to hypothetical, permanent changes in the brain and central nervous system, was discussed in depth but there were difficulties over interpretation of single and repeated episodes of HPNS and what, if any, the permanent effects might be. The significance of EEG abnormalities and consequent action in the absence of symptoms posed difficulties. If the EEG changes were regarded as trivial could it be assumed that there were trivial and reversible changes in the central nervous system? Other factors such as hypothermia, subject variation and a past history of damage to the CNS unrelated to diving, are difficult to evaluate in this context. Overt neurological decompression sickness following decompression from shallow air diving seems to most diving doctors to be by far the most important problem.

There are at present limitations to the use of neuropsychiatric testing to detect minor brain damage and, where brain damage is known to be present, a wide range of ability may still exist. The use of chamber dives in association with such tests has been advocated as having predictive value in the selection of subjects for operational deep diving.

The widest agreement appears to have been reached over the need for standards of assessment of the central nervous system, neuropsychiatric testing, and the pathology of CNS damage to be developed and the need to seek more refined methods of investigation. Because of the lack of hard evidence of actual neurological damage from deep diving and the need for more data, there seemed to be no case for limiting deep diving, but rather the requirement to monitor even more carefully those taking part in it, and to develop more effective means of doing so.

CHROMOSOME OBSERVATIONS IN DIVERS

A survey of the chromosomal changes in the T-lymphocytes of divers and a study of the genetic effects of pressure and gas mixtures on isolated blood cells (lymphocytes) have been reported recently from Aberdeen University.[14] In 153 divers (77 air; 76 heliox) there was an increased frequency of tetraploid cells (not a significant finding as regards health) and of chromosome type aberrations which were twice as common as in controls. This excess was in a few damaged cells in only 5% of the divers. It is not known why these cells are produced and they do not correlate with any of a number of factors considered including gas mixture, smoking or alcohol. It is difficult at this stage to judge how important these findings are and it is interesting that the experiments on isolated lymphocytes did not produce damaged cells.

HEARING AND LUNG FUNCTION IN DIVERS

There are two other potential long-term hazards which could involve cumulative changes, and over a long period of repeated diving lead to permanent disability. These are noise induced hearing loss, and the effects of saturation diving on lung function. Where work such as rock drilling under water or the use of high pressure water jets is carried out the noise level may be as high as 175 dB(A) and temporary threshold hearing loss may occur[15] in spite of attenuation of the sound pressure by the diving helmet. This suggests that methods of protecting divers from excessive sound pressure are needed and that audiometry is an essential part of medical examination. Measurements of lung function in seven divers[16] who had spent an average of 12 days in saturation at 290–300 m showed an increase in vital capacity of 5.4% but a decrease in transfer factor of 9% which lasted for some months: in one man the transfer factor did not return to normal until 8 months later. The full significance of these changes has to await further observations and follow up. But it is clear that respiratory function also needs to be monitored in relation to the diving activity presently being undertaken.

In conclusion it may be of interest to mention a mortality study of 2000 divers identified from the Decompression Sickness Central Registry records in the University of Newcastle upon

Tyne Social Security records of those men who were flagged in 1979 and so far we have knowledge of 19 deaths. In seven men, death was due to trauma of some kind, unassociated with diving activities. Four men died from illnesses which do not appear to have any connection with diving. Eight deaths were directly due to diving, four of them given as drowning and the other four due to diving accidents. So far there is no indication of any particular long-term risk from diving as such but obviously it is too early yet to do other then wait and see.

REFERENCES

1. The Diving Operations Special Regulations (1960). HMSO, London.
2. British Medical Association, Scottish Council Report of the working party on the medical implications of oil related industry (1975). BMA Scottish House, 7 Drumsheugh Gardens, Edinburgh, EH3 7QP
3. J. Hallenbeck, 'Central Nervous System' In *Workshop on Long-term Health Hazards of Diving*, Commission of the European Communities, Mines Safety and Health Commission, Luxembourg (1978).
4. W. P. Trowbridge, D. N. Walder and R. I. McCallum. *'An Estimate of the Age and Length of Experience of North Sea Commercial Divers at Retirement'* (1982) (Decompression Sickness Central Registry, University of Newcastle upon Tyne, England).
5. Decompression Sickness Central Registry, 'Aseptic bone necrosis in commercial divers', *Lancet* **2**, 384–388 (1981).
6. D. Hunter, *The Diseases of Occupations*, 5th edition, English Universities Press, London (1975) p. 805.
7. D. H. Elliott and E. P. Kindwall, *The Physiology and Medicine of Diving*, 3rd edition, (Ed. P. B. Bennett and D. H. Elliott), Baillière Tindall, London (1982) p. 466.
8. J. Wolkiewiez, 'Accidents Graves de la Plongée Amateur', *Proceedings of Symposium on Decompression Sickness*, Cambridge (1981) North Sea Medical Centre, Great Yarmouth.
9. F. L. Mastaglia, R. I. McCallum and D. N. Walder, 'Myelopathy associated with Decompression Sickness — a report of 6 cases', *Clin. and Exp. Neurology* (1983).
10. A. C. Palmer, I. M. Calder, R. I. McCallum and F. L. Mastaglia, 'Spinal Cord Degeneration in a case of "recovered" spinal decompression sickness', *Brit. Med. J.* 283, 888 (1981).
11. I. | Ròszahegyi, Neurological damage following decompression. In *Decompression of Compressed air workers in Civil Engineering*. (Ed. R. I. McCallum) Oriel Press, Newcastle upon Tyne (1967).
12. P. B. Bennett, The High Pressure Nervous Syndrome in Man. In *The Physiology and Medicine of Diving*, 3rd edition (Ed. P. B. Bennett and D. H. Elliott) Baillière Tindall, London (1982) p. 262.
13. The Long Term Neurological Consequences of Deep Diving. *EUBS/NPD Workshop*, Stavanger, 1983 (to be published).
14. D. P. Fox, 'Chromosome aberrations in divers', *Undersea Medical Research*, (in press).

15. O. I. Molvaer and T. Gjestland 'Hearing damage risk to divers operating noisy tools underwater', *Scand. J. Work Environ. Health* **7**, 263–270 (1981).

16. I. S. Davey, J. E. Cotes, D. J. Chinn and J. W. Reed, 'Does diving exposure induce airflow obstruction', *Clin. Sci.* **65**, 480 (1984).

14

Transfer Under Pressure: A Re-evaluation

Dr P. B. James, Woltson Institute

In 1974, International Underwater Contractors obtained a contract to dive for Conoco on the semi-submersible, Venture 1. At that time, Conoco had their headquarters in Dundee and Mr Andre Galerne, President of the diving company, was concerned about the morbidity and mortality in the North Sea. It had achieved an unenviable reputation, with deaths averaging about ten a year. Most of this was related to Northern sector experience, where the depths are in excess of 80 metres. Even bounce diving in such depths, requires prolonged decompression times and Mr Galerne was quick to point out that these are remote and hostile waters. Most of the activity at this time was in the drilling phase and it was argued that the Industry would soon move on to the construction phase, where the possibility of illness and trauma under pressure would increase. Saturation diving techniques would also involve prolonged decompression times.

His action, which deserves enormous respect, was to commission the South West Research Institute in San Antonio to build two titanium chambers. A small chamber, essentially a hyperbaric stretcher and a larger, two-person chamber, which would remain in the helicopter, together would allow divers to be transported from a facility offshore to a shore-based chamber. Although there was some delay in the commissioning of these chambers, they eventually appeared towards the end of 1977 and a successful demonstration, which many said was impossible, took place in 1978.

Although attitudes to medical evacuation have changed over the years, Mr Galerne's concern with safety was very laudable and he was prepared to actually finance his ideas. However, by 1978, many changes had taken place in the Industry. Systems had improved enough to actually allow divers to stand up fully in the chamber. The surface support crews became better trained, and much more experienced. There was also an accumulation of experience in the actual, as distinct from the theoretical, problems, because a great deal of underwater construction had taken place and many thousands of man-hours in saturation had been undertaken.

In 1978, I gave a presentation on the transport of casualties under pressure at the British Medical Association Conference in Aviemore. I said that I was not, at that time, aware of a case of near-drowning in bell diving where a diver had subsequently required intensive care, although it was clearly possible for the management of such a patient to be undertaken in the diving system offshore. In April 1984, such an incident happened aboard the Dundee Kingsnorth. There was an underwater explosion associated with a cutting operation, which shattered the visor of the helmet, ruptured the diver's eardrums and gave him a right-sided pneumothorax. During the rescue, he undoubtedly inhaled water. Last week, the bellman, Neil Wiggins, received the Frank Dearman Award for his superb recovery. He resuscitated the diver in the bell, but in the chamber the diver did require intensive care and a diver medic, who equally deserved an award, Ian Gray-Taylor, assisted his ventilation, gave him an intravenous infusion and generally looked after him in a most professional manner. Some 39 hours into the incident, which had been marked by a steady improvement in the diver's condition, with every vital sign improving, the chest X-ray arrived onshore. It showed what had been suspected; a pneumothorax and bilateral, pulmonary oedema. Some physicians were very much in favour of a transfer under pressure at this stage. However the patient was stable and conscious, with all the vital signs improving. It is necessary in this position, to draw up arguments for and against transfer. The main argument in favour of transport was that it would allow the diver to be placed in chamber onshore, in reasonable close proximity to hospital services. It was suggested that the patient still might need ventilation, because of sudden deterioration, and ventilation might be better achieved in a shore-based chamber. However, on reflection, it was really a rather difficult point to accept, because we do not know how to sedate a patient adequately at 150 metres, or indeed, whether a ventilator would work adequately.

In the against column was a very obvious factor. This man was still in considerable pain and distress, although stable, and the pain and the increased level of anxiety associated with transport could well have destablized him. There was the likelihood of a very extended period of, perhaps, 8 or 9 hours, with the casualty in a small chamber. The physician, who, at the time, had not been in a helium and oxygen saturation, would have required fairly rapid compression to 150 metres. This would probably have induced the changes of the high pressure nervous syndrome rendering the physician ineffective. The suggestion to transport under pressure came as a result of a discussion at about 1600 hours on a Friday afternoon. It is difficult to imagine a worse time to contemplate mobilizing resources. Had the decision been taken to mobilize at this point, because of the theoretical risk of rapid deterioration in the patient's condition, we would have been faced with a night transfer, with all the additional problems entailed.

The Archilles' Heel of the system is the helicopter, because failure of the aircraft means almost certain death for both the diver and the physician in the helicopter chamber. It is difficult to imagine the helicopter remaining afloat in a heavy sea, carrying these chambers. Large flotation bags could be fixed to the helicopter chamber for such an eventuality, but would require considerable development. The risks of this transfer could not be justified in view of the stabilization of the patient's condition and the fact that we had a physician offshore, together with a team of very competent fellow divers, who would have been able to ventilate the patient with a bag and mask arrangement should it have proved necessary.

The shore-based facility of the helicopter transport system has certain limitations, being a large, single lock chamber some distance from a hospital. However it is very difficult for any facility to keep on 24-hour stand-by, a high level of readiness and a high level of training when, in reality, it is so rarely necessary.

In view of these comments the question must therefore be raised. Is there a place for medical transport under pressure? Perhaps 8 or 9 years ago it could have been defended fairly rigorously, but now we can state, quite categorically, that there is no place for the acute transfer of a patient. It is, therefore, in my opinion, not necessary to have 24-hour stand-by as an emergency system. If the transport under pressure system has a role in emergencies, it could be much more strongly argued that it is relevant to amateur diving. For example, some of the problems which occur in the Orkneys and Shetlands, but there is a difficulty in maintaining any facility on a 24-hour stand-by and maintaining a team of extremely expensive people for amateur use.

On 21 November 1981, there was a storm in the North Sea and Transworld 58 broke all connections with the sea bed, together with the Sedco Phillips SS which, together, had a total of about 10 men under pressure. After the incident it was learned that the lifeboat chamber had been mobilized during this event on board the Sedco Phillips SS. The lifeboat chamber was originally tested on deck on a calm day, and it did actually float, but conditions on that day were rather different. Certainly, it would have been impossible to have used the crane. Those involved in mobilizing the chamber on this occasion possibly thought that if the Sedco Phillips SS collided with the platform and sank, the lifeboat chamber would just float gently clear. It is difficult to believe that it would do so. There are medical aspects of this particular problem and it is necessary in such circumstances to imagine personal involvement in this situation. The one voice that is almost never heard in a forum such as this, or at any of the other meetings that take place, is the active working diver — he is too busy earning his livelihood. Certainly, my preference would be to stay within the main structure and bolted on to the system in the event of a capsize. In such a situation, it is wise to remember that seasickness can kill quite rapidly. The life support is already provided within the hyperbaric lifeboat, and it would keep men alive in the water even if the chamber remained bolted on to the system.

These are very large structures, and in one acute capsize which has already taken place, the Alexander Kjelland, the vessel did not sink. Had there been divers in saturation in a system on this vessel, they probably would have survived. Experience of surface vessels in difficulties is that often those who stay with the vessel survive. When the Drill Master went aground during a tow, down the Norwegian coast, those who stayed on board were perfectly safe but several were killed in the lifeboat evacuation. In the case of a diving system, where divers are under pressure, there is a crucial difference for they are in a vessel which can be made water-tight for any depth in relation to the location. Diving chambers, like submarines, can be engineered to provide life-support, and a 48-hours capability would not be too difficult. Diving support vessels may be a special case, but I think the same suggestions may be relevant. In the Southern sector some agencies are requiring a hyperbaric evacuation system for a vessel which can only sink in a maximum depth of 40 metres of water, where it would clearly be better to provide in-chamber life support. It is not easy to lose a large vessel in about 40 metres of water.

Perhaps it may be fairly suggested that there are greater priorities to be faced than transport under pressure, and that efforts should be directed at some of the continuing problems. For example, single aperture bells where, if the divers become unconscious, it is impossible to get inside. Diver to surface communications are often poor and still impair the diver's working ability. In an emergency they may present real problems. In a recent incident, we had a perfect example of how technology has gone backwards. A tape of the incident containing crucial information, which could be of great benefit to everyone in the industry, is unintelligible. Earlier incident recordings in the same company were made using higher quality equipment and were comparatively easy to understand.